T0335688

Network Radar Countermeasure Systems

Qiuxi Jiang

Network Radar Countermeasure Systems

Integrating Radar and Radar Countermeasures

Qiuxi Jiang
Electronic Engineering Institute
Hefei, China

ISBN 978-3-662-48469-2   ISBN 978-3-662-48471-5   (eBook)
DOI 10.1007/978-3-662-48471-5

Library of Congress Control Number: 2015958881

Springer Heidelberg New York Dordrecht London
© National Defense Industry Press, Beijing and Springer-Verlag Berlin Heidelberg 2016
This work is subject to copyright. All rights are reserved by the Publishers, whether the whole or part of the material is concerned, specifically the rights of translation, reprinting, reuse of illustrations, recitation, broadcasting, reproduction on microfilms or in any other physical way, and transmission or information storage and retrieval, electronic adaptation, computer software, or by similar or dissimilar methodology now known or hereafter developed.
The use of general descriptive names, registered names, trademarks, service marks, etc. in this publication does not imply, even in the absence of a specific statement, that such names are exempt from the relevant protective laws and regulations and therefore free for general use.
The publishers, the authors and the editors are safe to assume that the advice and information in this book are believed to be true and accurate at the date of publication. Neither the publishers nor the authors or the editors give a warranty, express or implied, with respect to the material contained herein or for any errors or omissions that may have been made.

Printed on acid-free paper

Springer-Verlag GmbH Berlin Heidelberg is part of Springer Science+Business Media
(www.springer.com)

Foreword

In today's world, with the tide of change becoming more and more vigorous, the speed of innovation also accelerates. Particularly in the unknown and fierce military field, we need the spirit of innovation and call for innovative achievements. We are delighted to see the book *Network Radar Countermeasure Systems* introducing the concept of networking in the information age to radar countermeasures equipment systems, which is not only a scientific exploration of theoretical innovation, but also a successful work containing the spirit of innovation.

Radar and radar countermeasure, which form the important foundation of modern national defense and are significant symbols of the manifestation of defense capabilities and levels, are playing an increasingly important role in modern warfare. Since the use of radar from the 1930s, the battle of the radars has never stopped, and even intensified. This book is driven by these demands, following the concept of integrated development of the information age, to achieve the unions of passive and active, jamming and detection, reconnaissance and jamming, which is the major change in radar countermeasure systems in our country. Its approach, including concepts such as network, full-band coverage, transmitting and receiving remote deployment, distributed collaboration, network-based control, information sharing, and data fusion, not only contains a wide range of technological innovation, but also encompasses many new thoughts and ideas. What is more valuable is that the authors uses a lot of quantitative calculation and simulation data, to analyze and verify new ideas and new theories in depth, which not only increase the scientific network radar countermeasure system theory's credibility, but also allow access to a solid foundation of actual application.

Theory comes from practice and also guides practice, which can only be tested and improved in practice. This puts forward fresh ideas for new national comprehensive integrated air defense systems and provides guidance in China's national defense construction and military struggle preparation. I hope that this monograph can attract extensive attention and be put into engineering practice, into the reality of combat capability as soon as possible.

In view of this, I want to congratulate the publication of the book, hoping that it will inspire similar original books in the future.

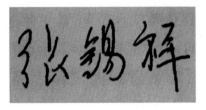

Chinese Academy of Engineering   Zhang Xixiang
June 1, 2010

Preface

The development and application of military radar has been the focus of both sides in electromagnetic warfare. Since radar began to be used for military purposes, radar countermeasures that have been developed have never stopped and are still appearing in an endless stream. In the future of information warfare, the employment of single radar and a single radar countermeasure device or simple combined use of multiple devices can no longer fulfill the needs of joint operations. Therefore, both radar and radar countermeasures are facing new challenges.

To be specific, radar is confronted with the following threats: (1) the active radiation of radar may cause straight attacks of anti-radiation weapons and electronic jamming; (2) radar detects targets via the reflection of electromagnetic waves, bringing about the emergence of stealth technologies that can absorb and reduce the electromagnetic reflection; (3) radar acquires target information through antenna beam scanning, which results in a lower space of silence and leaves a door wide open for aircraft to execute low-altitude penetration. It is these threats that impel the rapid development of radar in its functions and anti-jamming capabilities. For instance, a series of new types of radar, such as frequency agile radar, frequency diversity radar, pulse compression radar, repetition frequency (RF) change radar, phased array radar, synthetic aperture radar, Doppler radar, mono-pulse radar, bi-/multi-static radar, passive radar, non-cooperative passive detection radar, sparse array radar, and netted radar, have been developed to meet radar's functional and survival needs.

Given the emergence of various modern advanced radar systems and anti-jamming measures, radar anti-jamming technologies are presenting a strong trend towards integration. Accordingly, the traditional radar countermeasure equipments, which are based on a single station or a single method of countermeasure, also face severe challenges. An integrated use of radar countermeasures is the only way to make it effective.

Radar and radar countermeasure are two sides of the same coin. They are opposite to each other, yet depend on each other and also promote each other. In the network era, however, the integration of radar and radar countermeasure is not only an inevitable demand of modern information warfare, but also an irresistible trend of the development of electronic warfare (EW) systems and radar systems. The network era makes the demand and trend possible.

From the perspective of information acquisition, radar detection is an important means to acquire target information from land, air, sea, and space. But the information detected by radar simply includes the coordinates, tracks, and speeds of the targets, and are unable to satisfy the needs of information operations that require adequate information on the actual attributes, weapons systems, and information systems of the targets. Electronic reconnaissance can fill this gap. For example, the relevant information about target radiation sources can be acquired by electronic reconnaissance, the information about target coordinates, tracks, and speeds can be obtained by passive locating, and the information about target platforms and weapons systems can be obtained by electronic "fingerprint" identification. Hence, a comprehensive access to battlefield situation information can be gained through the integration of radar detection and electronic reconnaissance, namely, the integration of active and passive information.

From the perspective of complementarities, a radar system is faced with four severe challenges: anti-radiation weapons, stealth targets, low-altitude penetration, and electronic jamming. Various types of radar, such as netted radar, bi-static/multi-static radar, and sparse array radar, have been developed to cope with these challenges. Theoretically speaking, however, any single mode of radar system still has several problems that are difficult to solve. The main problem with radar countermeasure lies in the passivity and non-cooperation of information acquisition. That is to say, information acquisition and selection of jamming objects depend on the operation of target radiation sources. Therefore, if active emission and passive reception is integrated, the "advantages" of radar would be enhanced while the "disadvantages" of electronic reconnaissance would be avoided.

From the perspective of target attack, radar detection is the primary means of missile guidance and weapons control. It is an important guarantee for executing a "hard" precision strike on targets. Radar countermeasure, by contrast, is the primary means of conducting a "soft" attack on targets, which is an important guarantee for achieving supremacy of the information, the air, and the sea. From this perspective, therefore, radar detection and radar countermeasure must be integrated. In other words, hard and soft attacks must be integrated.

From the perspective of the developing trend of weapons, system of systems (SoS) combat is an outstanding characteristic of the high-tech war. For example, intelligence acquisition and target attack tend to be integrated in information wars. By integrating radar detection and radar countermeasure, once physical targets or electronic targets are identified, the attack system embedded in the SoS can be employed to carry out electronic jamming suppression and guide firepower to strike them.

From the perspective of the developing rules of networks, the integration of radar and radar countermeasure is an inevitable trend of the times. Network-based battlefield systems depending on highly developed network information technology would become the basic form of future wars, and the direction of future revolution has shifted from "sensor-oriented war" to "network-oriented war". Network-based battlefield systems would provide the commander with real-time and transparent spatial awareness via multi-sensor fusion.

So, the integration of radar and radar countermeasure is an inevitable product of the Internet age. The network radar countermeasure system we propose is one that incorporates both reconnaissance and jamming. It combines active detection and passive detection in an organic way so that the precision of tracking location could be greatly improved and the interruption of target information could be effectively avoided when all the enemy's radios remain silent. The system makes it possible to extract target information from multiple aspects, which not only contributes to the analysis and identification of threats, but also helps to fundamentally remove false alarms. Besides, it also ensures an accurate and reliable estimation of the threatening targets' properties, parameters, numbers, and locations within the airspace of the warning area. This could shorten the response time and facilitate quick jamming guidance so as to meet specific tactical needs.

The network radar countermeasure system has three basic working modes. The first is the active mode, in which the transmitter of the system can be used as a jammer. Another advantage of this mode is that, through sending friend-or-foe identification signals to targets, receiving stations can discriminate the targets according to the corresponding responses to the signals. That is to say, it is able to conduct secondary radar friend-or-foe identification. In this mode, the transmitter of the system can also function as a radar transmitter to send detection signals. With the reflection (scatter) echoes of the targets, detection and tracking can be completed. The second is the typical passive mode, in which radar signal reconnaissance and passive tracking location are carried out with the use of the signals emitted from target radiation sources. So, the network radar countermeasure system has functions of both electronic intelligence (ELINT) and electronic support measures (ESM), which could use the receiver to complete parameter measurement, reconnaissance, identification, and jamming parameter guidance of radiation source signals. The third mode is a combination of the active and passive modes, in which some of the transmitters and receivers of the system work together in the first mode, while other receivers work in the second mode, so that the system could achieve the best performance. The network radar countermeasure system operates based on network mode, which means that it implements the connection, management, coordination, and control of the transmitting and receiving stations through network protocols. In addition to performing functions like connecting, managing, coordinating, controlling, and timing all the network nodes, the central station or a certain designated receiving station must complete data fusion and processing, conduct target detection, location, identification, tracking, and track plotting, as well as provide military intelligence and electronic intelligence, such as target numbers, target properties, locations, threat levels, and electronic series, so as to make the detection area transparent and provide clear intelligence for electronic attack.

The network radar countermeasure system effectively solves the vital problem with passive radar, whose detection effectiveness is completely dependent on the radiation characteristics of targets. When targets remain radio-silent, passive radar would fail to work properly.

The network radar countermeasure system also effectively overcomes the weakness of non-cooperative passive radar. Despite the non-cooperation of external transmitters, the system can successfully detect and track targets in concealment.

I would like to express my deepest gratitude to some renowned academicians in China, such as Wang Xiaomo, Bao Zheng, Li Deyi, Zhang Lvqian, Mao Erke, and Zhang Xixiang, for their inspiring guidance and unfailing help. Without their illuminating direction, this book could not have reached its present form. Sincere thanks also go to Prof. He You and Prof. Tang Ziyue for their invaluable suggestions. Also, I especially want to thank academician Ling Yongshun for his critical review of the book.

It is worth noting that the book is a result of collective wisdom and combined efforts. I feel particularly indebted to all my leaders for their full recognition and great encouragement during the formative period of the concept. Also, I am grateful to my doctoral students, including Li Mingliang, An Zhen, Ding Feng, Shen Aiguo, Wang Bo, and Wang Zheng. For their vigorous support throughout the publishing process, I really owe countless thanks. Finally, I shall thank Prof. Qi Jianqing for her contribution to the structure arrangement and theoretical analysis of the book.

Network radar countermeasure systems is a newly proposed concept. Therefore, many theories and engineering technologies need to be further studied and applied to engineering practice. This book provides a new research topic and intends to start further discussions in that direction. Due to the limit of my ability, inadequacy is unavoidable in the book. Your criticism and suggestions would be highly appreciated.

Hefei, China   Qiuxi Jiang
August 28, 2010

Contents

Chapter 1
Introduction to Network Radar Countermeasure Systems

1.1 Introduction

Since the application of radar in the 1930s, its function has extended from initial target detection to today's utilization in warning detection, tracking guidance, navigation control, arms control, remote control, topographic mapping, weather forecasting, and other functions. Expansion of the working characteristics and its functions accelerate the uses of radar. Collision avoidance radar, Doppler radar, navigation radar, weather radar, surveying, mapping radar, and other civilian aspects have brought great convenience to people's daily lives, which makes them closely related to radar. Remote warning radar, target indication radar, missile guidance radar, navigation radar, weapons control radar, and target imaging radar are widely used in the military field. Radar has become the important foundation of modern warfare and homeland defense. More importantly, it has become a significant part of joint operations in the information age.

However, the development and application of military radar has become the focus of the two sides locked in battle. We can say that confrontations using radar have not been stopped since the day it was put to military use, and is becoming more intense. Active radar electromagnetic radiation leads to a direct attack from anti-radiation weapons and from electronic jamming, and radar detects the targets depending on the reflection of electromagnetic waves, which has led to the emergence of electromagnetic stealth technology, enabling the absorption and reflection of waves. Radar depends on antenna beam scanning to obtain target information, which produces blind space and prevents low-altitude penetration aircraft from working normally. However, the rapid development of radar is also attributed to the expansion of its function and the four challenging threats mentioned above, such as the emergence of frequency agile radar, frequency diversity radar, pulse radar, pulse repetition frequency (PRF) change radar, phased array radar, synthetic aperture radar, Doppler radar, monopoles radar, bi-/multi-radar, passive radar, non-cooperative passive radar,

© National Defense Industry Press, Beijing and Springer-Verlag Berlin Heidelberg 2016   1
Q. Jiang, *Network Radar Countermeasure Systems*,
DOI 10.1007/978-3-662-48471-5_1

sparse array radar, and networking radar, which resulted from its function and demands driven by everyday life.

The radar systems, such as frequency agile radar, frequency diversity radar, pulse compression radar, RF agile radar, phased array radar, synthetic aperture radar, Doppler radar, and monopoles radar, technically perform better in reconnaissance and anti-jamming tasks. But, as a single-station device, effective countermeasures and methods could always be found theoretically.

The transmitter and receiver are separated in bi-/multi-based radar systems, where the baseline range between the receiver and transmitter is of the same order of magnitude as the equivalent effective range. There are two kinds of bi-/multi-radars: the bi-static radar using accommodation transmitters and the composite bi-static radar using the joint work of single radars. The emitted radiation signal of a bi-static radar irradiates to the target, and the split receiver receives the scattered waves from the target and completes the detection and treatment. Each bi-static radar may appear to exhibit data redundancy; therefore, we can improve the positioning accuracy of bi-/multi- radar by combining and estimating this redundant data. So far, the typical bi-/multi- radars include the United States' Sanctuary air defense bi-static radar system, the tactical bi-static radar detection (TBIRD) system, bi-static alarm and control (BAC) system, multi-base measurement system (MMS), bi-static proximity warning system developed by the UK Plessey Company, bi-static radar experiment system studied at the University of London, Russia's Barrier radar, and so on. Especially, more and more bi-/multi-radar systems have been used in the homeland defense system in the United States, which are responsible for long-, middle-, and short-range strategic defense tasks. Multi-radar can be seen as a combination of a plurality of bi-static systems of transmitter stations. Each bi-static system firstly deals with the positioning processes independently, and then the results are transmitted to the central station for data integration, tracking, and other treatment. The key issue that should be solved by the bi-static system is the "three synchronization" problem of space, time, and phase, and solves the triangles problem formed by bi-static and objectives, which has a complex structure. The bi-static radar's resolution capability and accuracy are poorer than that of a single radar. What is more, the ability to distinguish between its sending and receiving baselines is almost lost. Due to the fact that the bi-/multi-radar's sending and receiving tasks are separated, antenna directivity only uses them in one direction separately, which leads to a great impact on side lobe clutter. The bi-static radar uses direction of arrival (DOA) direction angular positioning. Time difference of arrival (TDOA) positioning has some errors compared to the active and passive radar in positioning accuracy. What is more, a bi-/multi-mode base station arrangement has a greater impact on the detection area, and, therefore, there are many restrictions to its configurations and tactical uses.

Essentially, passive radar is the radar countermeasure reconnaissance system, which itself does not radiate electromagnetic waves, but obtains the target location and attributes by receiving electromagnetic waves radiated by the target. The development of passive radar began in the 1970s. The most famous products are the Tamara and VERA-E systems of the Czech company Tesla (now the ERA Company).

In addition, the Israeli-based EL/L-8300 and EL/L-8388 ground air defense electronic intelligence systems relied on a short time difference directional airborne system, which complete reconnaissance sorting and real-time tracking of multiple batches of air targets through the interception of airborne, shipboard, and ground-based radiation signals.

Russia and France also have similar equipment. China has equipped three Master baseline TDOA location system air passive detection equipment and four Master baseline TDOA location of a certain type of passive radar system. Passive radar itself does not radiate electromagnetic signals, receiving the radiated signal from target electronic equipment, compared with active radar, which has a long effect range, is concealed, and is not be easily found by others. In addition, in the working mechanisms of passive radar and positioning systems, the positioning methods are different from active radar, so the enemy's active electronic jamming equipment, which will become a passive radar's electronic jamming signal source, with be put to use to locate its position. Passive radar target detection is realized by receiving radiation from the target, regardless of its effective reflection cross-sectional area. So there is no essential difference in stealth aircraft and other aircraft. In addition, the passive radar can also discriminate the type of target and working conditions, etc., by intercepting radiation signal parameters. It can complete individual identification of targets through the identification of passive radar "fingerprint characteristics". However, the fatal problem of passive radar is that target detection depends totally on the presence of target electromagnetic radiation. If the target maintains radio silence status, passive radar would not work. In addition, other problems exist in passive radar, including: some disparities in detection and location precision compared with active radar; different forms of distribution corresponding to different location areas; the accuracy of low-altitude, long-range target location is poor, or even not possible at all. Passive radar only serves as a supplement and cooperative work for active radar.

Non-cooperative passive radar measures the TDOA and DOA of a television signal and FM broadcast signal's direct signal and uses the reflected signal from the target and Doppler shift to detect and locate the target. This positioning system has four advantages. First, it is similar to a bi-static radar or multi-stage radar system, and also works in the same wavelengths in its role of stealth target absorption as smaller FM radio and TV signals materials, it can deal with stealth targets effectively. Next, because the system itself does not emit electromagnetic energy, there is a good concealment, resisting radiation missiles effectively. Using the transmitted signal of commercial TV or FM radio stations, the system does not require expensive transmission equipment, which reduces costs. Finally, by working at low frequencies, it is not affected by small weather changes and is reliable, as well as yielding good system compatibility. But the downside is that the detection accuracy is not high enough, it cannot provide the data for precision attack weapons, and there will be a further study for chaff and other passive jamming on system performance. A prominent representative of non-cooperative passive radar is the Silent Sentry system from the US company Lockheed Martin, and the Defence Evaluation and Research Agency (DERA) is studying the use of television signals for use in a passive detection and location system. However, when using external radiation signals on the target for a passive detection and location system, there is a shortfall in military combat requirements on the positioning accuracy.

The sparse integrated aperture radar (simply SIAR) was proposed by the French ONERA (National Space Agency) in the late 1970s, which is a new type of metric wave, distributed array radar system, mainly used for long-range surveillance and tracking. The main feature of SIAR is that its antenna uses sparse arrays, non-directional emission, and the gain of its transmitting and receiving patterns can be increased by digital signal processing at the receiving end, so it can form multiple beams, observing multiple directions simultaneously. Because it does not use a traditional mechanical antenna to scan space beams, the beam formed by calculation can "peg" the target with coherent long periods of accumulation, which increases the radar detection and anti-threat capability. The SIAR adopts omnidirectional transmission with no difference in the main and side lobes in the air, so the receiver cannot obtain the radar information from its main beam and locate its general position like ordinary reconnaissance radar. The SIAR system determines that its transmitter and receiver antenna gain are relatively low; to achieve the effective detection of target echo, we necessarily need to adopt a coherent long period of accumulation. The radar transmitting power is small, so it has a better concealment and is difficult to be scouted. The SIAR is an active array radar, where the transmitting frequencies for the encoding and phase encoding of each array element are different and varied randomly, belonging to complex waveforms, so the opposite side cannot make use of signal processing means to obtain the transmitting beam, meaning that it is difficult to locate the radar position. But sparse arrays radar, generally using a level lineup, does not have a measurement capability of low altitude angle, and antenna arrays with large volumes are subject to certain restrictions in real-world applications.

The basic principles of the network radar is: multiple different systems, different frequency bands, different modes, and different polarization radars located suitably as the stations are linked into a network by the means of communication, and the central station leverages unified deployment and processing, so that the network becomes an organic whole. Target information obtained by each radar within the network (target track, flight path, etc.) is transmitted to the central station for processing, forming intelligence information within the target range of radar coverage, and according to the trend of operation changes to adjust the working status of radar, using the advantages of each radar, so that the detecting, locating, and tracking tasks will be completed in the entire coverage area. Radar networking is a technical measure with the characteristics of full-band, multi-institution, multi-overlapping technology which have been used in air defense and offense operations at home and abroad. Russia deployed the "rubber overshoes" Ballistic System, which is a typical monobasic radar network. A typical networking radar systems also include the CETAC air defense command center developed by the French company Thomson-CSF. Networking radar is restricted by current technology, still based on a single radar network, so the main problem is that the integration of data and information is based on the destination layer, while it is difficult to achieve signal processing and parameter level fusion, the data fusion degree is greatly restricted, and there are no obvious changes in the identification of the target compared with that of monobasic radar. The effect of monobasic radar has a great impact on networking radar.

The development of radar countermeasure depends on the development of radar. Regardless, the development of radar is in function extensions, performance improvements, or survival needs. Any changes in radar system and technology will have an impact on and challenge radar reconnaissance and jamming, even in regards to anti-radiation attacks. From reconnaissance, the multi-function wave, pulse code modulation, parameters and waveform agility, short pulse, and beam agility will bring great difficulties upon radar reconnaissance, signal sorting, and recognition. From the point of view of jamming, broadband pulse, coherent pulse, parameters agility, beam agility, bi-static, side lobe suppression, low side lobes, and multi-mode multi-function compatibility make the traditional radar jamming and other measures ineffective, and render single-station jamming equipment useless. From the point of view of anti-radiation of destruction attacks, difficulties caused by reconnaissance and radar jamming will pose challenges to the anti-radiation. In addition, there is a challenge for radiation source bait, which directly affects the combat effectiveness of anti-radiation weapons. However, these challenges brought about by radars greatly promote the development of radar countermeasure technology and its equipment. And the corresponding single-station countermeasure techniques and methods could be developed in some radar systems. In other systems, the problems could be solved by a system or a system of combat equipment. It is foreseeable that the development direction of the radar countermeasure equipment must be a systematic and comprehensive integration. The single-station equipment or the individual countermeasure is increasingly ineffective.

From the perspective of access to information, radar detection is an important means of access to space, air, sea, and ground targets. However, radar detection can obtain the coordinates, tracks, and velocity information of the target, which is difficult to meet the requirements of target attributes, operations, weapons systems, information systems, and other related information. Also, electronic reconnaissance is not the only source of obtaining radiation information from the target, but the coordinates, track, and velocity information of the target can also be obtained by passive reconnaissance. Through electronic "fingerprints" identification, relevant information about the target platform and weapons systems can be obtained. Therefore, to achieve the integration of radar and electronic reconnaissance, namely, active and passive intelligence integration, we can obtain comprehensive information on the battlefield situation.

From the complementary perspective, the radar system faces four challenges, anti-radiation weapons, stealth targets, low-altitude penetration, and electronic jamming, directed against the problem exposed by the radar, developing networking radar, dual/multi-radar, sparse arrays radar, etc., but in the single radar system there still exists insurmountable problems in principle. The problem that radar countermeasures face is the passive and non-cooperation of the access to information, namely, the choice of obtained information and jamming object depending on the target radiation processing. Therefore, to achieve the integration of active transmitter and passive reception, we can adopt the advantages of radar and avoid electronic surveillance disadvantages.

From the perspective of target attack, radar is the primary means of achieving arms control and missile guidance; to achieve high target accuracy is an important guarantee for "hard" combat. Radar countermeasure is the primary means to achieve the objective of a "soft" attack, which is to obtain control of information, control of the air, which is important to ensure the command of the sea. Therefore, from the perspective of target attacks, we must integrate radar detection and radar countermeasure, namely, soft and hard attack integration.

From the development of weapons, the system confrontation is the characteristic of the high-tech war, such as access to intelligence of the information war and strike integration. If radar detection and radar countermeasure integration can find an entity target or electronic target, we can use the electronic jamming system itself or boost firepower to destroy targets.

From the rule of network development, the integration of radar and radar countermeasure is an inevitable development of the times. Relying on a highly developed network of information technology, the network-based battle system is the basic form of war. The future direction of change is from "sensor-centric warfare" into a "network-centric warfare", depending on network-based battle systems, through a multi-sensor fusion commander's real-time, transparent spatial perception.

In summary, radar and radar countermeasure are two contradictory aspects, which are mutually antagonistic, interdependent, and mutually reinforcing. However, in today's Internet age, the integration of radar and radar confrontation is not only the inevitable demands of modern warfare and information warfare, but, also, the information age electronic warfare systems and radar systems are the inevitable trend of development. The Internet age makes such needs and trends possible, which leads to the radar network countermeasure system.

1.2 Overview of a Network Radar Countermeasure System

The phrase "network radar countermeasure system" refers to multiple transmitters, receivers, and hubs which are dispersed and deployed in different places. They are connected to be an organic whole by specific network protocols and equipment to achieve interoperability in the time domain, frequency domain, airspace, the completion of the target reconnaissance detection, intelligence gathering, identification and tracking, jamming suppression, fire and other comprehensive guide-integrated electronic information systems.

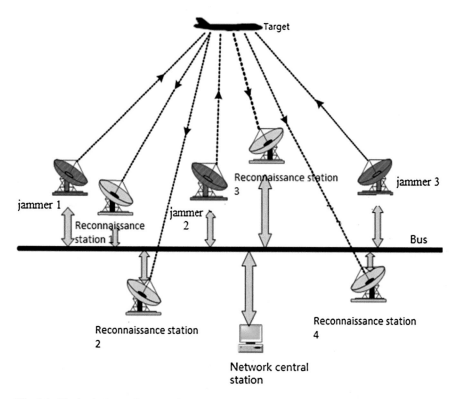

Fig. 1.1   The basic form of a network radar countermeasure system

1.2.1   *Working Principle of a Network Radar Countermeasure System*

Starting from the basic concept of the network radar countermeasure system, the basic components of the system consists of three parts. The first part is the network protocols and network equipment used for connection, communication, and linking. Secondly, the transmitting device (jamming station) is used for the beacon transmitter jamming signal, detection signal, and identification friend or foe (IFF) jamming signal. The third and final part is the interception, receiving, testing, processing, integration, identification, tracking, guidance, and control action of the reconnaissance receiving device (reconnaissance stations, including hubs), as shown in Fig. 1.1.

The basic principle of the system is that the decentralized and remotely deployed transmitting station transmits a signal and searches for specific regions under the control of the network central station, the decentralized and remotely deployed receivers intercept, test, receive, and process the electromagnetic signal transmitted and reflected by the target, and the interception, testing, receiving, processing, and transmission from the target form the basic data and information about the target

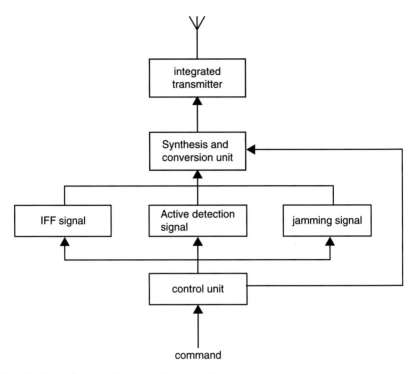

Fig. 1.2 Block diagram of the transmitter principle

from which some receivers of the central station can obtain target data and information through the network and perform data processing and information integration, complete information about the target and description, and achieve the goal of recognition and tracking, jamming suppression, and firepower boost.

The jamming station can perform three functions. The first is radar jamming signal transmission function of electronic jamming, the second is active radar signal transmission function, and the third is IFF signal transmission function. Each transmitting station scans the designated space in accordance with pre-programmed control instructions or hubs to narrow the (broad) beam irradiation detection airspace, to provide the receiving station with an illumination source.

The front and end of the transmitter's transmitting unit and a signal generating unit is compatible with the design needs. On the one hand, we need to meet the needs of the active radar transmitting signal waveform. On the other hand, we also need to take into account the signal waveform radar jamming and bandwidth needs. These requirements are shown in Fig. 1.2.

The receiver of the reconnaissance station receives the aerial target reflected echo signal or air target's radar signals, hubs, or any other receiving station, according to the target echo signals or radiation source or target, carries out the target detection, identification, location, and tracking, and, at the same time, gives its coordinate, tracking, velocity, and attribute information.

The receiving and signal processing subsystem focuses on the completion of radar echo signals, interception of radar emitter signal, detection, parameter extraction, target data recording, sorting and analysis of radiation source, identification, etc., and provides data support for the network radar countermeasure system data fusion and situation display subsystem. Among them, the passive reconnaissance mainly completes the radiation parameter measurements, signal sorting, target identification, and localization, and mainly provides information about the target attribute. The active probe mainly completes target detection, location, and tracking, and mainly provides state information about the target.

A receiving antenna uses a search system comprising a directed antenna. If using the phased array antenna system or multiple beam antennas, there may be increased probability of intercept and target data rate, but there are complexity and bandwidth constraints to consider as well.

The choice of receiver system must solve the problem of the requirement of a broadband receiver of passive reconnaissance, active detection of narrow-band matching (relatively narrow band), and compatibility the of receiver. The concept of integrated radio frequency is a key technology used by network radar countermeasure systems, but also results in the main difficulties of receiver antenna design and selection of the receiver system.

The signal processing section must follow the design principles of active and passive surveillance detection. Active signal processing mainly uses matched filtering, Doppler filtering, constant false alarm rate (CFAR), etc. to get the target azimuth information and range information, and carry out the reconnaissance fusion process according to the target position information from the different stations. Passive signal processing depends on the receiver's system and uses different systems with different processing algorithms and processing procedures, focusing on the full pulse data processing, signal sorting and matching, and signal recognition and targeting algorithms. There are several methods to choose the inter-pulse between signal analyses and signal processing pulse analysis. "Fingerprint" characteristic of the radar signal analysis is an effective way to achieve individual identification of the radar and platform. The radar signal arrival time will directly affect the measurement accuracy of the target location. A receiver and signal processing block diagram is shown in Fig. 1.3.

The central station consists of receivers and central control systems, in addition to the functions of reconnaissance work stations, and also has the function of a connection to each jamming station and reconnaissance station, of data transmission, communications, management, coordination, and control, according to the electromagnetic battlefield situation and task assignments, adjusting the working patterns of each node, pitch angle, azimuth, switch status, operating parameters, and completes the management and control of multiple modes of operation. The system operates in the active detection mode, where the network center controls the transmitting station number, work status, transmit power, transmit trigger, and other information, and transfers information to each surveillance station. When the system is in strong jamming or covert operation, the central station can control the network radar countermeasure system working in passive reconnaissance mode,

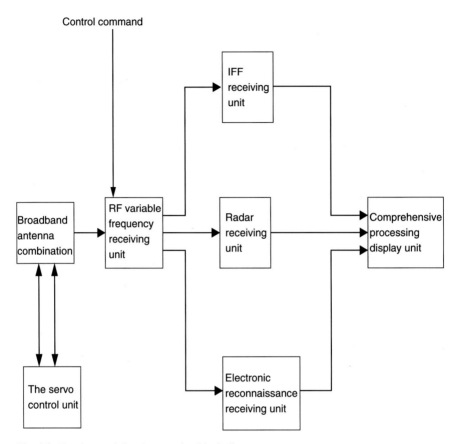

Fig. 1.3 Receiver and signal processing block diagram

using multi-station passive cross-location or method of station TDOA position measurements of the target, to complete scouted target technical parameters, processing, and recognition.

Data and information integration is the core function of the network central station. It communicates with each station to obtain the detection data integration and processing to complete the identification of targets, positioning, tracking and track plotting, and gives the number and attribute of the target, geographic location and the threat level, electronic sequences, etc., to achieve transparency in the detection region to provide direct information to the commanders.

Hubs provide an interface of other air defense units and combat systems, and support more agile combat establishment for the preparation of the implementation of a joint strike.

The system network mode allows users to access and exit the network any time, anywhere, and they can also access the network and share each network's node data and information. Therefore, the system has good flexibility and scalability.

1.2.2   Working Mode of the Network Radar Countermeasure System

The working principles of the network radar countermeasure system show that the system can work in active mode, passive mode, and active/passive integration mode. These will be covered in the following sections.

1.2.2.1   Active Mode

The active operating modes comprise two cases. The first is IFF and electronic jamming pattern, and the second is the active detection mode.

IFF and electronic jamming pattern: The working mode and parameters of the system transmitting station can be pre-programmed or controlled by the control command network hub station. When the need arises for radiation jamming targets, the transmitting station system can work in electronic jamming patterns, transmitting signals to the radiation target. Its jamming parameters are guided by the completion of the central station or the receiving station. In this case, the system constitutes a distributed network jammer, and the distributed jamming signal may enter from the radar antenna main lobe. The jamming signal will not be affected by the low side lobe antenna, side lobe, and side lobe suppression, thus its jamming efficiency is several orders of magnitude higher compared with side lobe jamming. Secondly, distributed jammers are located in different regions of the airspace, which can form a multi-directional fan. The combination of such multi-directional fans can form a large area of repressive jamming.

Similarly, the system is controlled by the command transmitting station, and can transmit the enemy interrogation signal. The receiving station receives the response signal and completes the IFF functions, constituting cooperative IFF systems, as shown in Fig. 1.4.

When the system is in the active detection mode, each transmitter station launches a radiation detection signal into space. The frequency of the system can cover a wide band of radar frequencies, but the single transmitters do not need such a wide frequency range, as its working methods and operating parameters can be pre-programmed or controlled by the control command network hub station. Each station detects, by aerial reconnaissance target reflection (scattering), the target echo signal and the reconnaissance probe data are passed to the network hub station. The receiving station can receive more than just the target echo transmitting station; it can also be the targeting single station. The hubs for each station carry out reconnaissance detection data fusion and correlation processing, completing the identification of targets, positioning, and tracking.

In this operating mode, the system can use the one transmitting multiple receiving mode, the multiple transmitting one receiving mode, and the multiple transmitting multiple receiving mode to detect the target. The one transmitting multiple receiving mode is shown in Fig. 1.5a and the multiple transmitting one receiving mode is shown in Fig. 1.5b. The multiple transmitting multiple receiving mode is

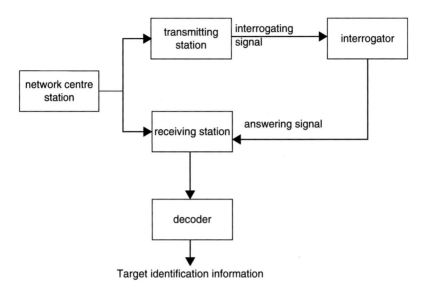

Fig. 1.4 Block diagram of a collaborative IFF system

Fig. 1.5 (**a**) One transmitting multiple receiving mode. (**b**) Multiple transmitting one receiving mode

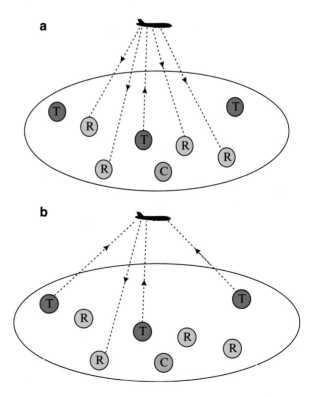

Fig. 1.6 Multiple
transmitting multiple
receiving mode

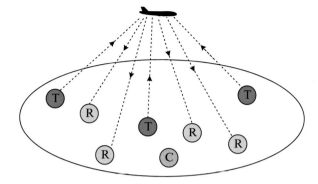

Fig. 1.7 Passive
working mode

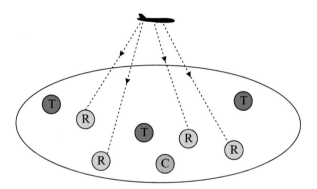

shown in Fig. 1.6. In the figures, T represents the transmitting station, R represents a receiving station, and C represents the hubs.

1.2.2.2 Passive Mode

When the system is in passive detection mode, each station receives target detection signals carrying the radiation source and performs analytical measurements for radar signal parameters for the signal recognition platform to identify and scout the results, which are then transmitted to the hubs. The hubs for each station carry out data and information fusion processing and correlation, and complete target identification, location, and tracking. Each station's reconnaissance data are also available to other station over the network, targeting multi-station functionality. A schematic of the passive detection mode of operation is shown in Fig. 1.7.

The electronic intelligence reconnaissance mode is a special case of the passive detection mode, but places more emphasis on the system's electronic intelligence reconnaissance capabilities. At this point, the focus is to obtain the system working parameters of enemy radiation and electronic order of battle.

1.2.2.3   Active Passive Integration Mode

When the system is in active and passive integration mode, the receiver and signal processing is compatible with both functions. Passive and active surveillance detect and complete the corresponding signal and data processing respectively, forming their own information on the target, and then performing the information fusion on the target layer. They can also work together in mutual support of information and guidance, to achieve the target location, tracking, recognition, and real-time guide jamming. Obviously, in this mode of operation, the identification of targets, tracking, and trajectory plotting is more accurate and provides better co-jamming.

Through rational design for the transmitter signal waveform, it is possible to achieve the integration of jamming and detection signals, This mode can jam targets and complete the reconnaissance detection.

1.3   Configuration of the Network Radar Countermeasure System

The network radar countermeasure system is an organic whole that connects multiple transmitters and multiple receivers through the Internet. Its detection performance depends on the configuration of each node and system size. Therefore, discussing the relationship between the configuration of the network radar countermeasure system and its performance is particularly important. From a geometric point of view, we will discuss three typical configurations: annular configuration, linear configuration, and zone configuration.

1.3.1   Annular Configuration

When the azimuth of the attacking target is uncertainty, or comprehensive protection for the target needs to be carried out, the detection area of the network radar countermeasure system needs to be comprehensive. In other words, the system should have a better ability of detection and location and try to reduce radar blind spots. At this time, the network radar system will use the annular configuration. The specific configuration is shown in Fig. 1.8.

The main advantage of the annular configuration is that the coverage of detection in all directions is basically the same, with the ability of target location, and it has a large coverage area. But its drawback is that it is unable to search for the directions of important targets and, therefore, continue with detection and location in those directions.

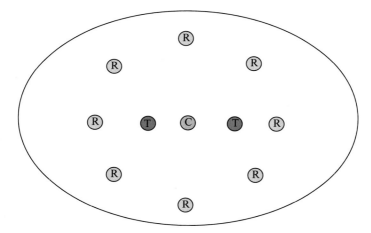

Fig. 1.8 Annular configuration

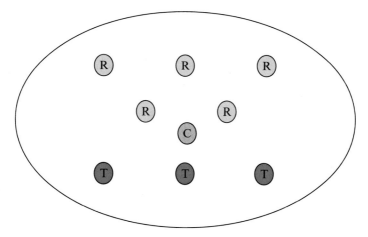

Fig. 1.9 Linear configuration

1.3.2   Linear Configuration

When the target's incoming direction is clear, the network radar countermeasure system only needs one direction of target detection and localization and the system can use the linear configuration. The linear configuration is shown in Fig. 1.9.

The linear configuration is also known as a directed configuration. Its main advantage is that the coverage of the frontal detection is the largest. The overlapping ranges of the detection areas are wide, target detection probability is high, and its anti-jamming ability is stronger than the other configurations. But there is still a problem. Due to its focus on the frontal direction, the detection capability of other directions is weakened, so the protection degree of freedom is small.

1.3.3 Zone Configuration

The network mode, especially the versatility of the hubs, makes the basic unit of multiple systems to be interconnected, and its detection area can, in principle, be an unlimited expansion. The so-called zone configuration means that it will be deployed in the detection area of each network connection radar countermeasure system of the basic unit, which forms an organic whole, to complete the target detection in the whole zone. In the zone configuration, each central station of the network radar countermeasure system not only accomplishes each transmitter and receiver's control and intelligence information extraction, but it can also serve as a central node in the network radar countermeasure system to transfer and share information. Two basic units of network radar countermeasure systems are interconnected to each other through a network interface of the central station, which forms a multi-level network radar countermeasure system. Each basic unit of network radar countermeasure systems in the zone configuration can be an annular configuration or linear configuration to maximize the operational efficiency of each network node according to the actual needs.

A particular note is that, in the zone configuration, according to the requirements of regional defense, the transmitting and receiving stations are staggered in configuration, in order to cover the entire area of defense. Therefore, in a wide range of homeland defenses, the zone configuration is the best option, as shown in Fig. 1.10.

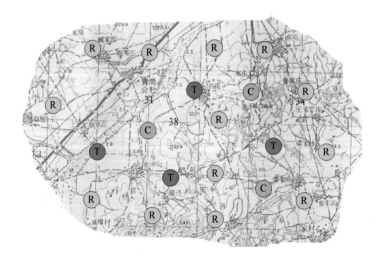

Fig. 1.10 Zone configuration of network radar countermeasure systems

1.4 Performance of Network Radar Countermeasure Systems

Reconnaissance coverage, jamming suppression area, and target resolution are the main functions of the system. In the following, we will discuss the performances of network radar countermeasure systems in the active mode, the passive mode, and the integrated active-passive mode.

1.4.1 Reconnaissance Detection Area of the Active Mode

The detection equation of a traditional mono-static radar is as follows:

$$P_r = \frac{P_t G^2 \lambda^2 \sigma}{(4\pi)^3 R^4 L} \tag{1.1}$$

In formula (1.1), P_r is the echo power received by the radar receiver, P_t is the transmitted power of the radar, G is the gain of the radar, λ is the wavelength, σ is the scattering cross-sectional area of the target, R is the range between the target and the radar, and L is the loss coefficient of all radar parts.

The power P_r received by the radar must exceed the minimum detectable signal power $S_{i\min}$ in order for the radar to find the target. If they are equal, we can only get the maximum range $R_{\max}$ where the radar detects the target, because beyond this range, the power P_r received by the radar is further reduced, thus failing to detect the target reliably.

The maximum range of the target is:

$$R_{\max} = \left[\frac{P_t G^2 \lambda^2 \sigma}{(4\pi)^3 S_{i\min} L} \right]^{1/4} \tag{1.2}$$

To detect the target under background noise, the receiver's output should amplify the signal to a sufficient value and, more importantly, the output signal to noise ratio (SNR) S_0/N_0 should reach the required value. In general, the quality of terminal signal detection depends on the SNR.

$$S_{i\,\min} = kT_0 B_n F_n \left(\frac{S_0}{N_0} \right)_{\min} \tag{1.3}$$

In the above formula, k is Boltzmann's constant, $k = 1.38 \times 10^{-23} J/K$, T_0 is the standard room temperature (generally, $T_0 = 290K$), B_n is the noise bandwidth of the receiver, and F_n is the noise coefficient of the receiver.

By substituting formula (1.2) into (1.3), we can obtain the following formula:

$$R_{max} = \left[\frac{P_t G^2 \lambda^2 \sigma}{(4\pi)^3 k T_0 B_n F_n \left(\frac{S_0}{N_0}\right)_{min} L} \right]^{1/4} \tag{1.4}$$

According to the traditional mono-static radar equation, when given a minimum detectable SNR $(S_0/N_0)_{min}$ of the output, we can calculate the maximum detection range of the radar and then obtain the power range of the radar.

Since the system's transmitting and receiving stations are deployed in different places in a decentralized manner, this makes its power range different from that of the mono-static radar. The power range of the mono-static radar is always a sphere with a radius of the maximum detection range, while the reconnaissance range of the network radar countermeasure system varies with the structure of the system.

1.4.1.1   Equation of Reconnaissance Detection

The probability density function of Gaussian noise, when output through the envelope detector, is:

$$p_0(r) = \frac{r}{\sigma^2} \exp\left(-\frac{r^2}{2\sigma^2}\right) \qquad r \geq 0 \tag{1.5}$$

where r represents the amplitude value of the noise envelope of the detector output.

When the sinusoidal signal with amplitude A is input to the envelope detector together with the Gaussian noise, the probability density function of the output of the envelope is:

$$p_A(r) = \frac{r}{\sigma^2} \exp\left(-\frac{r^2 + A^2}{2\sigma^2}\right) I_0\left(\frac{rA}{\sigma^2}\right) \qquad r \geq 0 \tag{1.6}$$

In formula (1.6), r represents the amplitude value of the output signal of the detector plus the noise envelope.

Suppose that the signals received by the receiving station are independent of each other, and the hubs use the ratio test criteria. Then, the test statistic is as follows:

$$T = \sum_i \ln\{p_A(r_i)/p_0(r_i)\} \tag{1.7}$$

In formula (1.7), i represents ith transceiver unit in the network radar countermeasure system.

If A is not a constant, but a Rayleigh random quantity with average amplitude σ_A, then the test statistic will be:

$$T = \sum_i \left(\ln \frac{\sigma_i^2}{\sigma_i^2 + \sigma_A^2} + \left(\frac{\sigma_A^2 r_i^2}{2\sigma_i^2 (\sigma_i^2 + \sigma_A^2)} \right) \right) \tag{1.8}$$

If $\sigma_A^2 = a\sigma_i^2$, then we can obtain the following:

$$T = \sum_i \left[ar_i^2 / \left(2(1+a)\sigma_i^2 \right) - \ln(1+a) \right] \tag{1.9}$$

By removing all the proportion parameters, the test statistic can be simplified as:

$$T = \sum_i r_i^2 / \sigma_i^2 \tag{1.10}$$

When the system is in active mode, we can assume that the system is completely coherent, and each receiving station can receive the echo signals emitted by all the transmitting stations in the system. At this time, the system can be viewed as being composed of $N \times M$ (N represents the number of transmitters and M represents the number of receivers) transceiver units and its reconnaissance detection range equation can be expressed as follows:

$$D_A = \sum_{i=1}^{M} \sum_{j=1}^{N} \frac{P_{t_i} G_{t_i} G_{r_j} \sigma_{ij} \lambda_i^2}{(4\pi)^3 k T_0 B_{n_j} F_{n_j} R_{ti}^2 R_{rj}^2 L_{ij}} \tag{1.11}$$

In formula (1.11), D_A is the active mode detection factor, defined as the minimum output SNR needed by detecting the target signal and:

P_{t_i}: Transmitting power of the ith transmitting station;
G_{t_i}: Antenna gain of the ith transmitting station;
G_{r_j}: Antenna gain of the ith receiving station;
λ_i: Wavelength of the signal transmitted by the ith transmitting station;
σ_{ij}: Radar cross-section (RCS) of the target relative to the ith transmitting station and the jth receiving station;
k: Boltzmann's constant;
T_0: Standard room temperature, generally 290 K;
B_{n_j}: Bandwidth of noise;
F_{n_j}: Noise coefficient;
R_{ti}: Range between the ith transmitting station and the target;
R_{rj}: Range between the jth receiving station and the target;
L_{ij}: System losses between the ith transmitting station and the jth receiving station.

When $N = M = 1$, the above formula is the equation of a bi-static radar; when the transmitting and receiving systems are integrated, i.e., $R_t = R_r$, then the formula is the equation of a mono-static radar.

Let us consider a simple case, in which the parameters of each transceiver unit are the same and the system works in an active mode. Then, the equation of the reconnaissance detection range can be expressed as follows:

$$D_A = \frac{P_t G_t G_r \sigma \lambda^2}{(4\pi)^3 kT_0 B_n F_n L} \sum_{i=1}^{M} \sum_{j=1}^{N} \frac{1}{R_{ti}^2 R_{rj}^2} \tag{1.12}$$

From formula (1.12), we can see that the equations of the reconnaissance detection range are closely related to the system structure (deployment location and number of transmitters and receivers), leading to the power range of the system also being affected by its system structure. In the following, we will discuss the relationship between the system power range and the system structure through simulation.

1.4.1.2 Simulation of the Reconnaissance Detection Area

Assuming that the system consists of two transmitting and two receiving stations, i.e., $M = N = 2$, and the parameters of the two transmitting stations are exactly the same, with specific parameters as follows: $P_t = 10\,\text{kW}, G_t = 40\,\text{dB}$, and wavelength of the transmitted signal $\lambda = 0.9\,\text{m}$. Then the parameters of the two receiving stations are also exactly the same, with specific parameters: $G_r = 20\,\text{dB}$, $F_n = 3\,\text{dB}, B_n = 9\,\text{MHz}$, radar cross-section of the target $\sigma = 10\,\text{m}^2$, system losses $L = 10\,\text{dB}$, and detection factor of the system operating in active mode $D_A = 13\,\text{dB}$.

To illustrate the advantages of the reconnaissance detection range performance of the system, we compare the power range of the network radar countermeasure system with that of a mono-static radar and the radar netting under the condition of the same parameters.

First, we compare the performance of the reconnaissance detection range with that of the mono-static radar. Assuming that the network radar countermeasure system consists of two transmitting and two receiving stations, in order to compare the reconnaissance detection range performance of the two kinds of radars objectively, we assume that the transmitting power of the mono-static radar is $P_t = 20\,\text{kW}, G_t = G_r = 30\,\text{dB}$, and the other parameters are the same as the network radar countermeasure system.

From the equation of the mono-static radar, we will obtain the maximum value as follows: $R_{\text{max}} = 48.85\,\text{km}$.

In Fig. 1.11, "*" is a mono-static radar and the shaded parts are the power range of the mono-static radar. By calculation, the two-dimensional power range of the mono-static radar in the figure accounts for 18.74 % of the total area. (The total area of the diagram is $200 \times 200\,\text{km}^2$.)

Consider a special system deployment. Assuming that the two transmitting stations and the two receiving stations of the system are put together, and the location is the same as the mono-static radar, and we mark it as deployment o. 1.

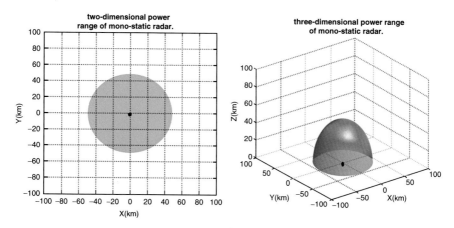

Fig. 1.11 Two-dimensional and three-dimensional power range of mono-static radar

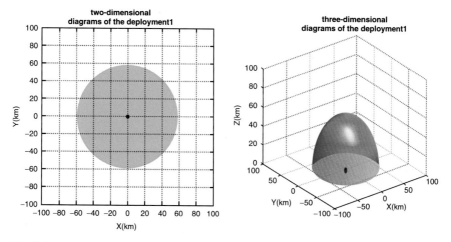

Fig. 1.12 Two-dimensional and three-dimensional diagrams of deployment No. 1

Through simulation, we obtain the two-dimensional and three-dimensional diagrams of the power range under deployment No. 1 as shown in Fig. 1.12.

By calculation, we find that the two-dimensional power range of the system accounts for 26.50 % of the total area in the case of deployment No. 1. Although the total transmitting power of the system is the same as that of the mono-static radar, its power range become larger. This is because each of the receiving stations can receive the echo signal transmitted by all of the transmitting stations, which increases the total power of the receiver and amplifies its power range.

Consider putting transmitting station 1 and receiving station 1 together, with the location $TR_1 : x_1 = 0\,\mathrm{km}, y_1 = 30\,\mathrm{km}, z_1 = 0\,\mathrm{km}$, and transmitting station 2 and receiving station 2 together, with the location

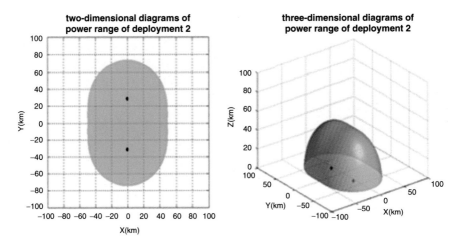

Fig. 1.13 Two-dimensional and three-dimensional diagrams of the power range of deployment No. 2

$TR_2 : x_2 = 0\,\text{km}, y_2 = -30\,\text{km}, z_2 = 0\,\text{km}$, and mark it as deployment No. 2. Through simulation, we obtain the two-dimensional and three-dimensional diagrams of the power range under deployment No. 2 as shown in Fig. 1.13.

By calculation, we find that the two-dimensional power range of the system accounts for 31.27 % of the total area in the case of deployment No. 2. As compared with deployment No. 1, the two-dimensional power range of deployment No. 2 is increased, but the height of its three-dimensional power range is decreased. This shows that, when the total transmitting power is certain, there is a price to be paid for increasing the two-dimensional power range.

Now, let us compare the performance of the reconnaissance detection range of radar netting in the case of deployment No. 2. Assuming that radar netting consists of two mono-static radars and their position is the same as that in deployment No. 1, then the transmitting power of each mono-static radar is $P_t = 10\,\text{kW}$, $G_t = G_r = 30\,\text{dB}$, and the other parameters are the same as those of the network radar countermeasure system. The power range of radar netting is equal to the overlay of all the power ranges of the mono-static radars, i.e.:

$$V_{\text{netted}} = \bigcup_{i=1}^{N} V_i \tag{1.13}$$

where V_{netted} is the power range of radar netting and V_i is the power range of the ith mono-static radar in the radar netting.

From Fig. 1.14, we can glean the information that the two-dimensional power range of the radar netting accounts for 24.36 % of the area in the diagram. Under the same conditions, the power range of the system is greater than that of radar netting.

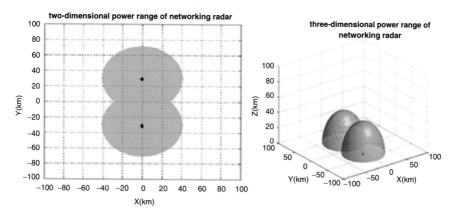

Fig. 1.14 Two-dimensional and three-dimensional power range of radar netting

Similarly, each receiving station of the system can receive the echo of all the signals emitted by the transmitting stations, resulting in an increase of the receiving signal power and its power range.

Consider a more general deployment of system; that is to say, the transmitting and receiving stations are deployed dispersedly in different locations. Assume that the coordinates of the stations are as follows:

$$T_1 : tx_1 = -40\,\text{km}, ty_1 = -40\,\text{km}, tz_1 = 0\,\text{km},$$
$$T_2 : tx_2 = 40\,\text{km}, ty_2 = -40\,\text{km}, tz_2 = 0\,\text{km},$$
$$R_1 : rx_1 = -40\,\text{km}, ry_1 = 40\,\text{km}, rz_1 = 0\,\text{km},$$
$$R_2 : rx_2 = 40\,\text{km}, ry_2 = 40\,\text{km}, rz_2 = 0\,\text{km},$$

marked as deployment No. 3 of the network radar countermeasure system. By simulation, we obtain the two-dimensional and three-dimensional diagrams of the power range under deployment No. 3 as shown in Fig. 1.15.

In Fig. 1.15, "#"stands for a transmitting station and "*"stands for a receiving station.

By calculation, we can see that the two-dimensional power range of the system accounts for 33.68 % of the total area in the case of deployment No. 3, which indicates that proper deployment of transmitting and receiving stations can increase the two-dimensional power range, but the height of the three-dimensional power range of deployment No. 3 is decreased remarkably. It shows that increasing the two-dimensional power range has to pay a price when the total transmitting power is certain.

Take the case where one more transmitting station and receiving station is added with the coordinates $T_3 : tx_3 = 0\,\text{km}, ty_3 = -40\,\text{km}, tz_3 = 0\,\text{km}$ and $R_3 : rx_3 = 0\,\text{km}, ry_3 = 40\,\text{km}, rz_3 = 0\,\text{km}$, marked as deployment No. 4 of the network radar countermeasure system. By simulation, we obtain the

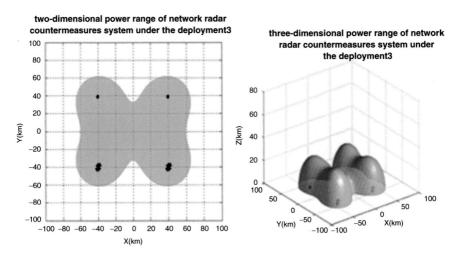

Fig. 1.15 Two-dimensional and three-dimensional power range of the network radar counter-measure system under deployment No. 3

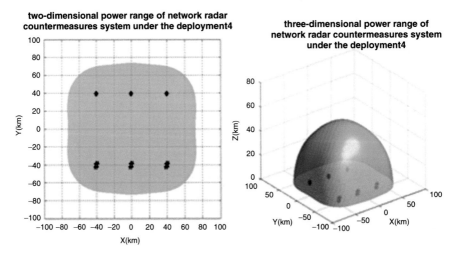

Fig. 1.16 Two-dimensional and three-dimensional power range of the network radar counter-measure system under deployment No. 4

two-dimensional and three-dimensional diagrams of the power range under deployment No. 4 as shown in Fig. 1.16.

From Fig. 1.16, we know that, by adding transmitting and receiving stations to the system, the power range can be further increased.

The above analysis of simulation shows that the reconnaissance detection range performance of a network radar countermeasure system has an obvious advantage over either mono-static radar or radar netting with the same transmitting power, and that proper deployment of transmitting and receiving stations in terms of location and number can improve its reconnaissance detection range. Therefore, we can reasonably distribute the stations in the network radar countermeasure system according to the specific monitoring area.

1.4.2 Reconnaissance Detection Area in Passive Mode

1.4.2.1 Equation of Reconnaissance Detection

Similarly, the range of reconnaissance detection in the passive working mode can be expressed as follows:

$$D_P = \sum_{j=1}^{N} \frac{P'_t G'_t G_{r_j} \lambda_p^2}{(4\pi)^2 k T_0 B_{n_j} F_{n_j} R_{rj}^2 L_j} \qquad (1.14)$$

In formula (1.14):

D_P: Passive mode detection factor;
P'_t: Transmitting power of the target radar;
G'_t: Antenna gain of the target radar;
λ_p: Transmitted signal wavelength of the target radar;
L_j: System losses of the jth receiving station;

The other parameters are the same as those in the active working mode.

When $N = 1$, formula (1.14) is a reconnaissance equation of the reconnaissance receiver.

In the same way, let us consider a simple case in which the parameters of each receiving station are the same. When the system works in the passive mode, the reconnaissance detection range equation can be expressed as follows:

$$D_P = \frac{P'_t G'_t G_r \lambda_p^2}{(4\pi)^2 k T_0 B_n F_n L} \sum_{j=1}^{N} \frac{1}{R_{rj}^2} \qquad (1.15)$$

From (1.15), we can see that the equation of the reconnaissance detection range in the passive mode is also closely related to its system structure, which means that the power range of the system is also affected by the system structure.

1.4.2.2 Simulation of the Reconnaissance Detection Area

Suppose that the specific parameters of each receiving station in the passive mode are the same as those in the active working mode, i.e., the specific parameters of

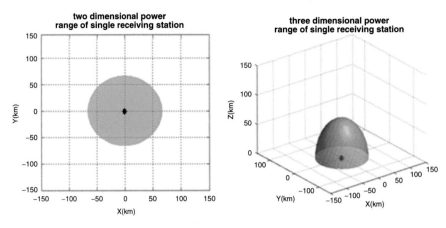

Fig. 1.17 The two- and three-dimensional power range of a single receiving station

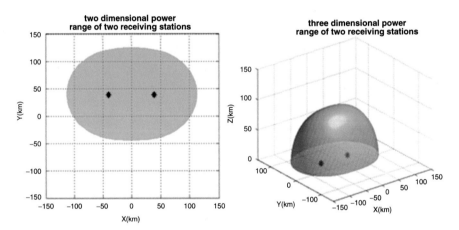

Fig. 1.18 The two- and three-dimensional power range of two receiving stations

target radiation are: $P_t' = 100\,\text{W}$, $G_t' = 10\,\text{dB}$, and $\lambda_p = 0.01\,\text{m}$. The system losses and detection factors are also the same as those in the active mode.

In the following, we shall analyze the reconnaissance detection range performance of one receiving station, two receiving stations, three receiving stations, and four receiving stations by simulation (namely $N = 1, 2, 3, 4$). In Figs. 1.17, 1.18, 1.19, and 1.20, "*" stands for the receiving station in the diagrams.

Coordinates of a single receiving station: $x = 0, y = 0, z = 0$;

Coordinates of two receiving stations:

$$x1 = -40, y1 = 40, z1 = 0; \quad x2 = 40, y2 = 40, z2 = 0;$$

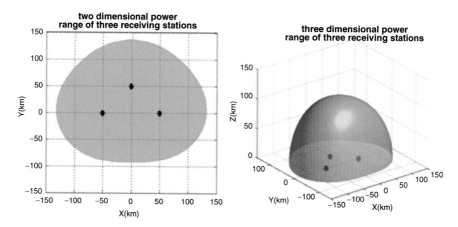

Fig. 1.19   The two- and three-dimensional power range of three receiving stations

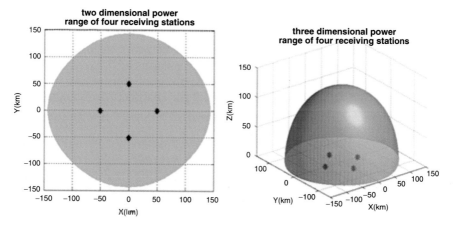

Fig. 1.20   The two- and three-dimensional power range of four receiving stations

Coordinates of three receiving stations:

$$x1 = 0, y1 = 50, z1 = 0; \quad x2 = -50, y2 = 0, z2 = 0;$$
$$x3 = 50, y3 = 0, z3 = 0;$$

Coordinates of four receiving stations:

$$x1 = 0, y1 = 50, z1 = 0; \quad x2 = -50, y2 = 0, z2 = 0;$$
$$x3 = 50, y3 = 0, z3 = 0; \quad x4 = 0, y4 = -50, z4 = 0;$$

From the results of simulation, we can see that, in the passive working mode, the power range of the system is directly related to the number and deployment of receiving stations, but has nothing to do with transmitting stations. Therefore, we can make a reasonable distribution of receiving stations in the network radar countermeasure system.

1.4.3   Reconnaissance Detection Area in the Integrated Active-Passive Mode

1.4.3.1   Equation of Reconnaissance Detection

Similarly, the reconnaissance detection range in the integrated active-passive mode can be expressed as:

$$D_H = \text{Max}\left[\sum_{i=1}^{M}\sum_{j=1}^{N}\frac{P_{t_i}G_{t_i}G_{r_j}\sigma_{ij}\lambda_i^2}{(4\pi)^3 kT_0 B_{n_j}F_{n_j}R_{ti}^2 R_{rj}^2 L_{ij}}, \sum_{j=1}^{N}\frac{P_t'G_t'G_{r_j}\lambda_p^2}{(4\pi)^2 kT_0 B_{n_j}F_{n_j}R_{rj}^2 L_j}\right] \quad (1.16)$$

In formula (1.16), D_H is the detection factor of the integrated active-passive mode, and other parameters are the same as those in the active and passive modes. Max$[a, b]$ refers to the choice of the larger value among a and b.

Similarly, we consider a simple case in which the parameters of each transmitting and receiving station are the same. Then, the reconnaissance detection range of the network radar countermeasure system in the integrated active-passive mode can be expressed as:

$$D_H = \text{Max}\left[\frac{P_t G_t G_r \sigma\lambda^2}{(4\pi)^3 kT_0 B_n F_n L}\sum_{i=1}^{M}\sum_{j=1}^{N}\frac{1}{R_{ti}^2 R_{rj}^2}, \frac{P_t'G_t'G_r\lambda_p^2}{(4\pi)^2 kT_0 B_n F_n L}\sum_{j=1}^{N}\frac{1}{R_{rj}^2}\right] \quad (1.17)$$

From formula (1.17), the equation of the reconnaissance detection range in the integrated active-passive mode is also closely related to the system structure, which means that the power range of the system is also affected by the system structure.

1.4.3.2   Simulation of the Reconnaissance Detection Area

Assume that the specific parameters of simulation in the integrated active-passive mode are the same as those in the active and passive working modes ($M = N = 2$), and its deployment is the same as in deployment No. 3 of the active mode. Its two- and three-dimensional power range is shown in Fig. 1.21.

Comparing Fig. 1.21 with Figs. 1.15 and 1.18, and we can see that the power range in the integrated active-passive mode is the overlay of those in the active and passive modes, meaning that the power range of the system is maximal.

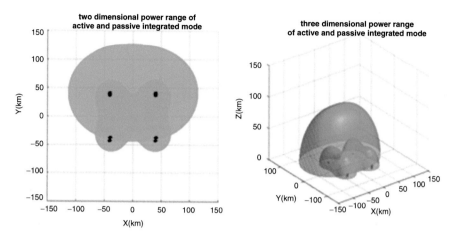

Fig. 1.21 The two- and three-dimensional power range in the integrated active-passive mode

1.4.4 Reconnaissance Detection Area in Jamming Conditions

1.4.4.1 Calculation of the Detection Area of the Network Radar Countermeasure System in Jamming Conditions

From its working principle, we know that blanket jamming destroys or reduces the target detection capability of the reconnaissance detection system so as to increase the observation error. If the network radar countermeasure system is suppressed by the enemy's blanket jamming, observational data of the system on the target are related to the strength of jamming.

When the jammer carries out a blanket jamming on the network radar countermeasure system, we can only implement side lobe jamming on its receiver, since it is difficult to ascertain the location of the reconnaissance stations. Compared with conventional aligned jamming on a known receiver's location, the ratio of the jamming signal power to the target echo signal power (jam-to-signal ratio) is smaller.

The jamming signal power P_C and the target echo signal power P_R are expressed, respectively, as follows:

$$P_C = \frac{P_j G_j' \lambda^2 G_r}{(4\pi)^2 R_R^2} \tag{1.18}$$

$$P_R = \frac{P_t G_t G_r \lambda^2 \sigma}{(4\pi)^3 R_T^2 R_R^2} \tag{1.19}$$

where:

P_j: Transmitting power of the jammer;
G'_j: Side lobe gain of the jammer;
G_r: Receiver gain of the reconnaissance station;
P_t: Transmitting power of the transmitter in the system;
G_t: Gain of the transmitter;
σ: Target cross-section;
R_R: Range between the target and the receiver of the reconnaissance station;
R_T: Range between the target and the transmitter station.

When formula (1.18) is divided by formula (1.19), we can calculate the jam-to-signal ratio of a reconnaissance unit made up of a pair comprising the cooperating transmitting station and reconnaissance station in the network radar countermeasure system.

$$K_{jC} = \frac{P_C}{P_R} = \frac{4\pi P_j G'_j}{P_t G_t \sigma} R_T^2 \tag{1.20}$$

In formula (1.20), G'_j usually takes the empirical formula:

$$\frac{G'_j}{G_j} = k \left(\frac{\theta_{0.5}}{\theta} \right)^2$$

In this formula, G_j is the main lobe gain of the jammer, $\theta_{0.5}$ is the half power beam width, θ is the intersection angle between the side lobe of the reconnaissance station and the main lobe of the jammer, and k is the coefficient, generally 0.04–0.10.

In some areas, when the jam-to-signal ratio K_{jC} is greater than the blanket coefficient K_{jC} (typical value 3), the reconnaissance unit will not find the target. These areas are called "jamming areas". In other areas, where the jam-to-signal ratio K_{jC} is less than K_j, the reconnaissance system can detect the target. The areas consisting of these ranges are called "exposure areas". Of course, the exposure areas must also meet the condition of the maximum detection range. Therefore, the detection areas of the reconnaissance unit consist of the solution set of regions of the following inequalities:

$$\begin{cases} \dfrac{P_C}{P_R} \leq K_j \\ \dfrac{P_t G_t G'_r \lambda^2 \sigma}{(4\pi)^3 R_T^2 R_R^2} \geq P_{Rr\min} \end{cases} \tag{1.21}$$

In formula (1.21), $P_{Rr\min}$ is the minimum detectable power of the receiver station.

Suppose that there are M transmitting subsystems and N receiver subsystems in the system, consisting of a reconnaissance unit $M \times N$. When calculating based on the method described above, we obtain the detection area Ω_i of each reconnaissance unit $R_i (i = 1, 2, \cdots, M \times N)$ and the detection area of the entire network radar countermeasure system $\Omega = \bigcup\limits_{i=1}^{M \times N} \Omega_i$ under single jammer conditions.

If multiple jammers, under the guidance of their reconnaissance system, carry out jamming actions on the network radar countermeasure system simultaneously, the jammers can only aim at the transmitting stations and the jamming signals cannot enter from the side lobe of the receiving station, as the receiving station of the network radar countermeasure system does not radiate the electromagnetic waves, thus the effect of multiple jammers being not ideal.

The power of the jamming signal can be superimposed at the receiver station. According to relevant principles, we can obtain the jam-to-signal ratio of a single reconnaissance unit K_{jC} as:

$$K_{jC} = \frac{\sum\limits_{i=1}^{m} P_{Ci}}{P_R} = \frac{4\pi R_T^2}{P_t G_t \sigma} \sum\limits_{i=1}^{m} P_{ji} G_{ji}' \tag{1.22}$$

In formula (1.22), m is the number of jammers, P_{ji} is the transmitting power of the ith jammer, G_{ji}' is the side lobe gain of the ith jammer, P_t is the transmitting power of the transmitter station in the system, G_t is the gain of the transmitter, and σ is the cross-section of the target.

Thus, the detection area of the reconnaissance unit is the solution set of the following inequalities Ω:

$$\begin{cases} \dfrac{\sum\limits_{i=1}^{m} P_{Ci}}{P_R} \leq K_j \\[2ex] \dfrac{P_t G_t G_r' \lambda^2 \sigma}{(4\pi)^3 R_{T1}^2 R_{R1}^2} \geq P_{Rr\min} \\[2ex] \qquad \vdots \qquad\qquad\qquad (i,j = 1,2,\cdots,m) \\[2ex] \dfrac{P_t G_t G_r' \lambda^2 \sigma}{(4\pi)^3 R_{Ti}^2 R_{Rj}^2} \geq P_{Rr\min} \\[2ex] \qquad \vdots \\[2ex] \dfrac{P_t G_t G_r' \lambda^2 \sigma}{(4\pi)^3 R_{Tm}^2 R_{Rm}^2} \geq P_{Rr\min} \end{cases} \tag{1.23}$$

where R_{Ri} is the range from the target to the ith reconnaissance receiver and R_{Ti} is the range from the target to the ith transmitting station in the system.

Assume that the system has M transmitting subsystems and N receiver subsystems, consisting of a reconnaissance unit $M \times N$. When calculating based on the method described above, we can obtain the detection area Ω_i of each reconnaissance unit as $R_i(i = 1,2,\cdots,M \times N)$, and then the detection area of the entire network radar countermeasure system $\Omega = \bigcup\limits_{i=1}^{M \times N} \Omega_i$ under single jammer conditions.

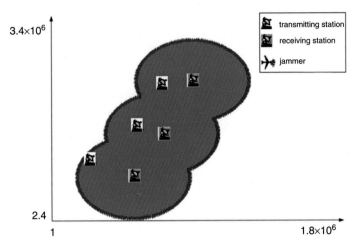

Fig. 1.22 The linearly deployed detection area without being jammed

1.4.4.2   Simulation of the Detection Area of the Network Radar Countermeasure System Under Jamming Conditions

Assume that the system configures three transmitting stations and three receiving stations with the following parameters: $P_t = 3\,\text{MW}$ $G_t = G_r = 35\,\text{dB}$ $L = 12\,\text{dB}$. Take, for example, jamming from one jammer with the parameters: $P_j = 5\,\text{KW}$ $G_{j_r} = 20\,\text{dB}$.

Experiment 1 The system is linearly deployed, with the transmitting stations at $[1119581, 2698706]$, $[1290201, 2876869]$, $[1385403, 3084006]$, the receiving stations at $[1281260, 2612247]$, $[1389092, 2838651]$, $[1503915, 3112636]$, and the jammer at $[1754606, 2971581]$.

By simulation calculation, we find that, when it is linearly deployed without being jammed, the detection area is somewhat similar to an ellipse, as shown in Fig. 1.22. Under single jammer conditions, the detection area is practically unchanged, as shown in Fig. 1.23.

Experiment 2 The system is circularly deployed, with the transmitting stations at $[1113635, 2762269]$, $[1315000, 3052752]$, $[1410661, 2797465]$, the receiving stations at $[1271303, 2790930]$, $[1343836, 2935324]$, $[1253078, 2892887]$, and the jammer at $[1754606, 2971581]$.

By simulation calculation, we find that, when it is circularly deployed without being jammed, the detection area is somewhat similar to a flower-shaped region, as shown in Fig. 1.24. Under single jammer conditions, the system detection will produce two small notches facing the jammer, but there is no great change in the detection area, as shown in Fig. 1.25.

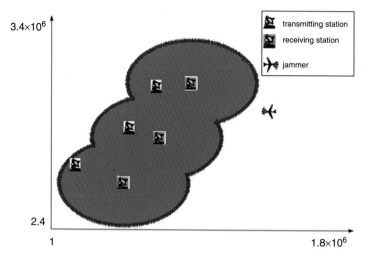

Fig. 1.23 The linearly deployed detection area under single jammer conditions

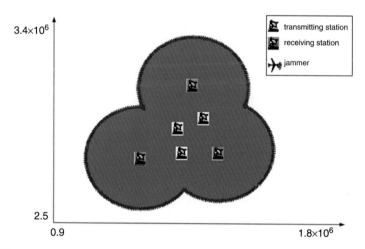

Fig. 1.24 The circularly deployed detection area without being jammed

Experiment 3 The system is linearly deployed, with the transmitting stations at [1119581, 2698706], [1290201, 2876869], [1385403, 3084006], the receiving stations at [1281260, 2612247], [1389092, 2838651], [1503915, 3112636], and the jammers at [1754606, 2971581], [1678825, 2744669], [1410661, 2797465].

By simulation calculation, we find that, when it is linearly deployed under three jammers conditions, the detection area is virtually unchanged because it is difficult for the jammers to aim at all the receivers in the system simultaneously, as is shown in Fig. 1.26.

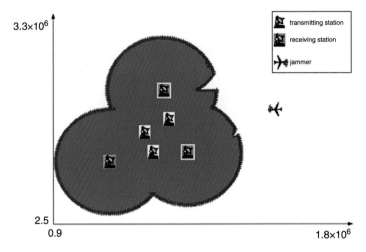

Fig. 1.25 The circularly deployed detection area under single jammer conditions

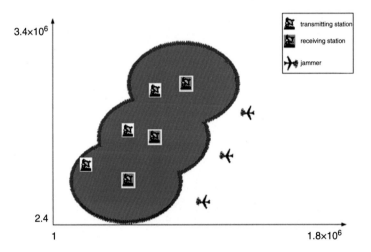

Fig. 1.26 The linearly deployed detection area under multiple jammers conditions

Experiment 4 The system is circularly deployed, with the transmitting stations at [1119581, 2698706], [1290201, 2876869], [1385403, 3084006], the receiving stations at [1281260, 2612247], [1389092, 2838651], [1503915, 3112636], and the jammers at [1754606, 2971581], [1678825, 2744669], [1410661, 2797465].

By simulation calculation, we find that, when it is circularly deployed under three jammers conditions, the system detection area will produce two small notches facing the jammers, which affects the system's detection performance to some extent, but the overall detection performance of the system is not decreased remarkably, as shown in Fig. 1.27.

Based on the above four groups of experiments, we can see that jammers have little impact on the detection performance of the whole system, be it linearly or

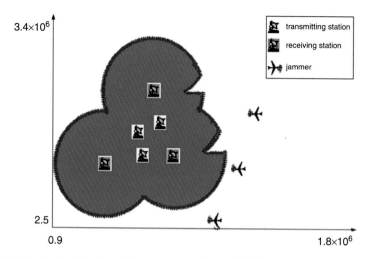

Fig. 1.27 The circularly deployed detection area under multiple jammers conditions

circularly deployed. In particular, when the system is linearly deployed, the detection area of the system is virtually unaffected by the jammers.

1.4.5   Jamming and Suppression Area of the Network Radar Countermeasure System

1.4.5.1   Jamming and Suppression Area

In the active mode, the central station guides the jamming station to suppress the enemy's airborne radar to defend its own goals. When the system protects the ground targets, there is a great difference in the formed effective jamming area due to the difference in the jamming power emitted from the subjamming systems and the position of deployment. Therefore, we need to start with the spatial and energy relationship between the jamming station, airborne radar, and the target to be detected in order to study the system's capability of suppressing the enemy's airborne radar.

Analysis of the Suppression Capability of a Single Jammer to a Single Airborne Radar

When jamming the enemy's fire-controlled radar, the jamming stations are usually deployed around the targets to be protected. In this case, the spatial relationship between the jamming station, the target, and the radar is as shown in Fig. 1.28:

The height of the bomber is H, the rectilinear range and the projection range from H to the protected area are R_t and D_t, respectively, and the rectilinear range and the projection range from the jamming station to the bomber are R_j and D_j, respectively.

In order to suppress the radar reliably and make it unable to find the target, we should suppress it when its antenna points at the edge of the target. Therefore, the

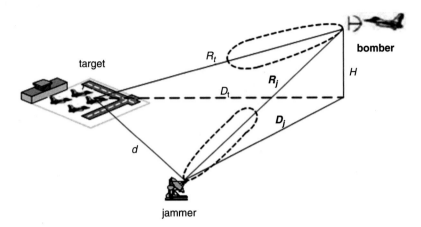

Fig. 1.28 The spatial relationship between the jamming station, the target, and the radar

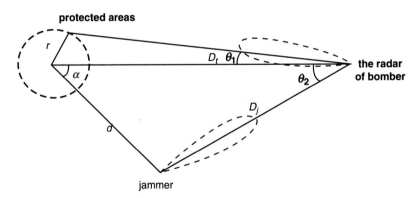

Fig. 1.29 The ground projection relationship between the jamming station, the target, and the radar

angle of the jamming signal deviating from the radar maximum gain direction should be $\theta_j = \theta_1 + \theta_2$, where θ_1 and θ_2 can be obtained from Fig. 1.29, as follows:

$$\theta_1 = \arcsin\frac{r}{D_t} \tag{1.24}$$

$$\theta_2 = \mathrm{arctg}\,\frac{d\sin\alpha}{D_t - d\cos\alpha} \tag{1.25}$$

The power of the jamming signal from the jammer received by the airborne radar is:

$$P_{rj} = \frac{P_j G_j G'_t(\theta)\lambda^2\gamma_j}{(4\pi)^2 R_j^2} \tag{1.26}$$

In formula (1.26):

P_j: Transmitting power of the jammer;
G_j: The jammer's antenna gain (in the direction of the main lobe);
R_j: Rectilinear range between the jammer and the radar;
γ_j: Polarization loss of the jamming signal to the radar antenna, generally 0.5;
$G_t'(\theta)$: Radar antenna gain in the main lobe direction of the jammer, generally calculated by the following formula:

$$
G_t'(\theta) = \begin{cases} G_t & 0 \leq \theta \leq \dfrac{\theta_{0.5}}{2} \\ K\left(\dfrac{\theta_{0.5}}{\theta}\right)^2 G_t & \dfrac{\theta_{0.5}}{2} \leq \theta \leq 90° \\ K\left(\dfrac{\theta_{0.5}}{90}\right)^2 G_t & \theta \geq 90° \end{cases} \tag{1.27}
$$

In formula (1.27):

G_t: Radar antenna gain in the main lobe direction;
$\theta_{0.5}$: Lobe width of the radar antenna;
K: Constant, generally $0.04 \sim 0.10$.

The ratio between the jamming power at the input of the radar receiver and the echo signal power, i.e., the jam-to-signal ratio is:

$$
\frac{J}{S} = \frac{P_{rj}}{P_{rs}} = \frac{P_j G_j}{P_t G_t} \cdot \frac{4\pi\gamma_j}{\sigma} \cdot \frac{R_t^4}{R_j^2} \cdot \frac{G_t'(\theta)}{G_t} \tag{1.28}
$$

In formula (1.28):

P_t: Transmitting power of the airborne radar;
G_t: Antenna gain of the airborne radar;
σ: RCS of the protected area.

As shown in formula (1.28), when the ratio of the jamming power and the echo signal power J/S is greater than or equal to the suppression coefficient K_j required by the power criteria, we can obtain the jamming equation as follows:

$$
\frac{P_j G_j}{P_t G_t} \cdot \frac{4\pi\gamma_j}{\sigma} \cdot \frac{R_t^4}{R_j^2} \cdot \frac{G_t'(\theta)}{G_t} \geq K_j \tag{1.29}
$$

Generally, $\theta_1 + \theta_2 \in \left(\frac{\theta_{0.5}}{2},\ 90°\right)$, so the gain of the radar antenna in the direction of the jammer is:

$$
G_t'(\theta_1 + \theta_2) = K\left(\frac{\theta_{0.5}}{\theta_1 + \theta_2}\right)^2 G_t \tag{1.30}
$$

Fig. 1.30 Schematic
diagram of a single jammer
deployment

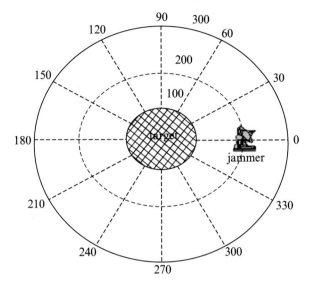

Substituting formula (1.29) into the jamming equation yields:

$$\frac{P_j G_j}{P_t G_t} \cdot \frac{4\pi \gamma_j}{\sigma} \cdot \frac{R_t^4}{R_j^2} \cdot \frac{G_t'}{G_t} \geq K_j \tag{1.31}$$

$A \triangleq \dfrac{P_t G_t \sigma K_j}{P_j G_j 4\pi \gamma_j K \theta_{0.5}^2}$, then:

$$\frac{R_t^4}{R_j^2} \cdot \frac{1}{(\theta_1 + \theta_2)^2} \geq A \tag{1.32}$$

$$f(D_t, \alpha) \equiv \frac{R_t^4}{R_j^2} \frac{1}{(\theta_1 + \theta_2)^2} = \frac{(D_t^2 + H^2)^2}{D_t^2 + H^2 + d^2 - 2D_t d \cos \alpha} \left[\arcsin \frac{r}{D_t} + \operatorname{arctg} \frac{d \sin \alpha}{D_t - d \cos \alpha} \right]^{-2} \tag{1.33}$$

Given $P_t G_t = 100 \times 10^6 \,\mathrm{W}$, $P_j G_j = 100 \times 10^3 \,\mathrm{W}$, $K_j = 5$, $K = 0.03$, $\sigma = 1 \times 10^4 \,\mathrm{m}^2$, $\gamma_j = 0.5$, $\theta_{0.5} = 3°$, $r = 100 \,\mathrm{m}$, $d = 400 \,\mathrm{m}$, and $H = 1000 \,\mathrm{m}$. If the jammer jams the airborne radar of an F16-C, the deployment of a single jammer is shown in Fig. 1.30.

By establishing a coordinate system poling with the target center, we can draw the suppression area of the single jammer to the airborne radar through formula (1.33), as shown in Fig. 1.31.

From formulas (1.31), (1.32), (1.33) and Fig. 1.31, we can make a conclusion as follows:

1. The main factor that affects the suppression area and the exposed area is A, which is determined by such parameters as the jammer, the target, and the radar.

Fig. 1.31 The suppression area of a single jammer jamming the airborne radar

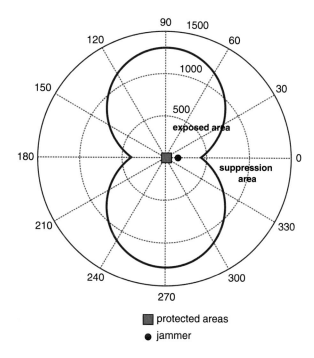

protected areas

● jammer

That is, the bigger the $P_j G_j$, the smaller the exposed area will be; the bigger the $P_j G_j$ and σ, the greater the exposed area.

2. The exposure radius is minimum and second to minimum in the $\alpha = 0°$ and $\alpha = 180°$ directions, with values of 399.9 m and 382.1 m, respectively. However, it is maximum when $\alpha = 90°$ and $\alpha = 270°$, with the exposure radius being 1300 m in both cases. As a result, when using jammers to defend targets, we should deploy them in the main direction of the enemy's attack, protruding or retreating a certain distance d, so that the exposure radius of the target in the main direction of the enemy's assaulting aircraft becomes the smallest.

3. The deployed position of the jammer, i.e., its distance to the target center d, has a major impact on the shape of the suppression area. When $d = 0$, i.e., the position of the jammer is in the center of the target, the exposed area is a circle. The bigger the d, the greater the exposure radius in the directions $\alpha = 90°$ and $\alpha = 270°$, the flat-longer the exposed area, and the greater the difference in the exposure radius in the directions $\alpha = 180°$ and $\alpha = 0°$. When d becomes the smallest, the exposure area approaches a circle. In tactical application, therefore, we should determine the size of d reasonably in order to reach the best jamming effect.

Performance Analysis of the System Suppressing a Single Airborne Radar

The system's multiple jammers carry out jamming on a single radar simultaneously under the guidance of the reconnaissance system. In this case, we can carry out the analysis based on the principle that the power of jamming signals can be superimposed in the radar receiver. According to this principle, we can obtain the total jamming power received by the radar as follows:

$$P_{rj} = \sum_i P_{rji} = \frac{G_t \lambda^2 \gamma_j K \theta_{0.5}^2}{(4\pi)^2} \sum_i \frac{P_{ji} G_{ji}}{R_{ji}^2 \theta_{ji}^2} \tag{1.34}$$

where $\theta_{ji} = \theta_{1i} + \theta_{2i}$.

The ratio between the jamming power received by the radar and the target echo power, i.e., the jam-to-signal ratio, is:

$$\frac{J}{S} = \frac{P_{rj}}{P_{rs}} = \frac{4\pi \gamma_j K \theta_{0.5}^2 R_t^4}{P_t G_t \sigma} \sum_i \frac{P_{ji} G_{ji}}{R_{ji}^2 \theta_{ji}^2} \tag{1.35}$$

Thus, we can arrive at the jamming equation of the radar as follows:

$$\sum_{i=1}^n \frac{P_{ji} G_{ji}}{R_{ji}^2 (\theta_{1i} + \theta_{2i})^2} \geq \frac{P_t G_t \sigma K_j}{4\pi \gamma_j K \theta_{0.5}^2 R_t^4} \tag{1.36}$$

In the case when we use three jammers with the same parameters to carry out jamming on the airborne radar of an F16-C, given $P_t G_t = 100 \times 10^6 \, W$, $P_j G_j = 100 \times 10^3 \, W$, $K_j = 5$, $K = 0.03$, $\sigma = 1 \times 10^4 \, m^2$, $\gamma_j = 0.5$, $\theta_{0.5} = 3°$, $r = 100 \, m$, $d = 400 \, m$, $H = 1000 \, m$, the three jammers are deployed as shown in Fig. 1.32.

Fig. 1.32 Schematic diagram of three jammers deployment

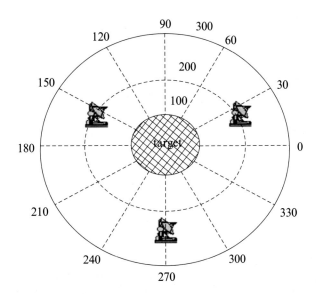

Fig. 1.33 The suppression area of three jammers jamming an F16-C bomber's radar

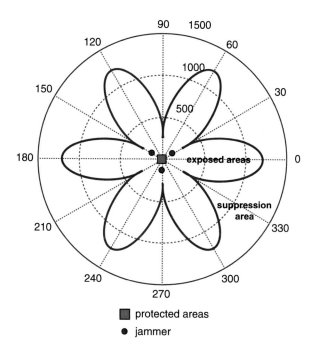

■ protected areas

● jammer

Then, we can use $R_t^2 = D_t^2 + H^2, R_{ji}^2 = D_{ji}^2 + H^2$ and formula (1.34) to calculate the exposed area and the suppression area of the radar when jammed by multiple jammers.

When the three jammers are deployed as shown in Fig. 1.33 to carry out jamming on the airborne radar, the exposure radiuses are minimum in directions $\alpha = 30°$, $\alpha = 90°$, $\alpha = 150°$, $\alpha = 210°$, $\alpha = 270°$, and $\alpha = 330°$, being about 264.4 m; the maximum occur in directions $\alpha = 0°$, $\alpha = 60°$, $\alpha = 120°$, $\alpha = 180°$, $\alpha = 240°$,and $\alpha = 300°$, being about 1230 m. We can see that, when multiple jammers carry out jamming on the airborne radar, the maximal exposure radius and the minimal exposure radius are remarkably smaller than those when using a single jammer to jam the airborne radar, thus being more effective in protecting the targets.

1.4.5.2   Simulation of Jamming and Suppression Area

Simulation 1 Jamming and suppression performance in jamming the airborne radar of early warning aircraft to cover own fighters to intercept the enemy's air raid aircraft fleet.

The system is operating in the active mode. If the transmitting station transmits a co-signal, the receiving station will receive the co-signal to locate the early warning aircraft actively; if the transmitting station is used as the jamming station, the

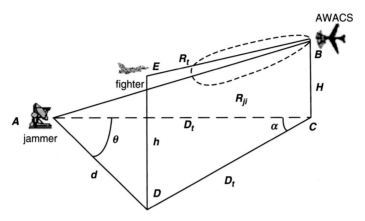

Fig. 1.34 The spatial relationship of the jamming station in covering the fighter

receiving station will covertly detect the radiation signal of the early warning aircraft and locate its position by dual-station or multi-station direction finding. Under the condition where the network radar countermeasure system can locate the early warning aircraft, the jamming station of the system can trace and jam the early warning aircraft.

Generally, the system is linearly or circularly deployed facing the incoming targets. When the system is jamming the airborne radar of the early warning aircraft to cover the fighters to intercept the air raid fleet, the spatial relationship between the early warning aircraft, the fighter, and the jamming station is as shown in Fig. 1.34.

Assuming that the early warning aircraft flies to and fro along the interior side of the battle line and considering the enemy's position in the shooting range of air-to-air missiles and the intercepting line of the fighter, there should be a certain distance between the patrolling flight path of the early warning aircraft and the battle line, called *discovery near boundary*, generally 80–100 km.

The length of the patrolling side of the early warning aircraft is:

$$S = \frac{1}{2}\left[\frac{2A - B}{tg\alpha} - \frac{A}{\sin \alpha} - \pi R + \sqrt{A^2 - B^2}\right] \qquad (1.37)$$

In formula (1.37), $tg\alpha = \frac{V_T}{V_A}$ is the velocity ratio, V_A (km/h) is the flight speed of the early warning aircraft, V_T (km/h) is the target velocity, A (km) is the radar detection horizontal range of the early warning aircraft to the specific target (with a certain RCS and flight altitude), B (km) is the discovery near boundary, R (km) is the turning radius of the early warning aircraft, and S (km) is the length of the patrolling side of the early warning aircraft.

Simulation 2 Simulated calculation of the exposure area of track-shaped airborne early warning radar.

Assume that the system has n jamming stations, then the jamming equation of the ith jamming station is:

$$\frac{4\pi\gamma_j}{\sigma} \cdot \frac{\left(D_t^2 + (H-h)^2\right)^2}{P_t G t_t^2} \cdot \frac{P_{ji}G_{ji}G_t(\theta_i)}{R_{ji}^2} \cdot \frac{\Delta f_r}{\Delta f_{ji}} = K_j \qquad (1.38)$$

Then, the jamming equation of the system is:

$$\frac{4\pi\gamma_j}{\sigma} \cdot \frac{\left(D_t^2 + (H-h)^2\right)^2}{P_t G t_t^2} \cdot \sum_{i=1}^{n} \frac{P_{ji}G_{ji}G_t(\theta_i)}{R_{ji}^2} \cdot \frac{\Delta f_r}{\Delta f_{ji}} = K_j \qquad (1.39)$$

In formula (1.39), n is the number of jammers, P_{ji} is the transmitting power of the ith jamming station, G_{ji} is the antenna gain of the ith jamming station, $P_{ji}G_{ji}$ is the equivalent power of the ith jammer, γ_j is the polarization loss of the jamming signal to the radar antenna, generally 0.5, P_t is the transmitting power of the airborne radar of the early warning aircraft, G_t is the airborne radar's antenna gain of the early warning aircraft, σ is the fighter's RCS, K_j is the suppression coefficient, $G_t(\theta_i)$ is the airborne radar's antenna gain of the early warning aircraft in the jamming direction, Δf_{ji} is the jamming signal's bandwidth of the ith jammer, and Δf_r is the bandwidth of the radar receiver.

Take the jamming radar of the early warning aircraft for example. There are, altogether, three jamming stations in the system, linearly deployed with an equal distance of 300 km. Given $P_t G t_t = 81\,\mathrm{dBW}$, $P_{j1}G_{j1} = P_{j2}G_{j2} = P_{j3}G_{j3} = 50\,\mathrm{dBW}$, $D_{j1} = 300\,\mathrm{km}$, $D_{j2} = D_{j3} = 424.2\,\mathrm{km}$, $K_j = 30$, $\frac{\Delta f_r}{\Delta f_{j1}} = \frac{\Delta f_r}{\Delta f_{j2}} = \frac{\Delta f_r}{\Delta f_{j2}} = 0.3$, and $\sigma = 10\,\mathrm{m}^2$, the exposure area of the jamming stations to the track-shaped airborne early warning radar is shown in Fig. 1.35.

From the simulation results, we can see that the jamming stations should be deployed linearly along the direction of the early warning aircraft movement when carrying out jamming on the early warning aircraft to cover the fighters of the air defense. Meanwhile, a proper interval should also be selected based on the flight path of the early warning aircraft, which should ensure that the angular range of the notch covers the intercepting direction of the fighters and that the covering of the main direction is emphasized. Based on the analysis of the effect of different parameters on the shape of the exposure area, the distance between the jammers and the early warning aircraft should be reduced as much as possible in order to obtain the smallest radius of the exposure area in the direction of the jamming stations. Therefore, in deploying jamming stations, we should move them forward a certain distance in the direction where the fighters carry out interception under the condition when the jamming stations are safe enough.

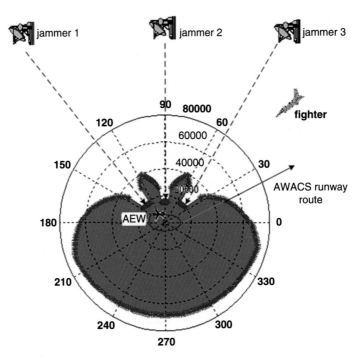

Fig. 1.35 The exposure area of the suppressing track-shaped airborne early warning radar

1.4.6 Fuzzy Function in the Mode of Reconnaissance Detection

When the system detects multiple targets, there is also the problem of resolution. Luckily, fuzzy function can not only describe the characteristics and the fuzziness of the resolution of the radar signals, but also the measuring accuracy and the clutter's inhibitory properties determined by the radar signals.

1.4.6.1   Expressions of Fuzzy Function

Assume that the system is fully coherent; each receiving station can receive the echo signals from all the transmitting stations in the system and extract the echo signals independently through the matching filter groups. Then, the difference between the echo signals of the two received targets A and B is:

$$\varepsilon_{\text{network}}^2 = \sum_{j=1}^{N}\sum_{i=1}^{M}\int_{-\infty}^{\infty} \eta_{ij}\left|s_i\left(t - \tau_{Aij}\right)e^{j2\pi f_{Aij}t} - s\left(t - \tau_{Bij}\right)e^{j2\pi f_{Bij}t}\right|^2 dt$$

$$= \sum_{j=1}^{N}\sum_{i=1}^{M}\left[\int_{-\infty}^{\infty}\eta_{ij}\left|s_i\left(t-\tau_{Aij}\right)\right|^2 dt + \int_{-\infty}^{\infty}\eta_{ij}\left|s_i\left(t-\tau_{B_{ij}}\right)\right|^2 dt\right] - \qquad (1.40)$$

$$2\text{Re}\left[\sum_{j=1}^{N}\sum_{i=1}^{M}\eta_{ij}\int_{-\infty}^{\infty}s_i\left(t-\tau_{Aij}\right)s_i^*\left(t-\tau_{Bij}\right)e^{j2\pi\left(f_{Aij}-f_{Bij}\right)t}dt\right]$$

In the formula, N is the number of receiving stations in the system, M is the number of transmitting stations, $s_i(t)$ is the complex envelope of the transmitting signal of the ith transmitting station, τ_{Aij} is the time used by the transmitting signal of the ith transmitting station to reach the jth receiving station via target A, τ_{Bij} is the time used by the transmitting signal of the ith transmitting station to reach the jth receiving station via target B, f_{Aij} is the Doppler shift used by the transmitting signal of the ith transmitting station to reach the jth receiving station via target A, f_{Bij} is the Doppler shift used by the transmitting signal of the ith transmitting station to reach the jth receiving station via target B, and η_{ij} is the weighted coefficient, defined as:

$$\eta_{ij} = \frac{P_{Rij}}{\text{Max}\left(P_{Rij}\right)} \qquad (1.41)$$

In formula (1.41), P_{Rij} is the echo power of the transmitting signals of the ith transmitting station received by the jth receiving station, the expression of which is:

$$P_{Rij} = \frac{P_{ti}G_{ti}G_{rj}\lambda_i^2\sigma_{ij}}{(4\pi)^3\left(R_{ti}R_{rj}\right)^2} \qquad (1.42)$$

In formula (1.42), P_{ti} is the transmitting power of the ith transmitting station, G_{ti} is the antenna gain of the ith transmitting station, G_{rj} is the antenna gain of the jth receiving station, λ_i is the wavelength of the transmitting signal of the ith transmitting station, σ_{ij} is the RCS of the target relative to the ith transmitting station and the jth receiving station, R_{ti} is the range between the ith transmitting station and the target, and R_{rj} is the range between the jth receiving station and the target.

Then, the fuzzy function of the system can be expressed as:

$$\chi_{\text{network}} = \sum_{j=1}^{N}\sum_{i=1}^{M}\eta_{ij}\int_{-\infty}^{\infty}s_i\left(t-\tau_{Aij}\right)s_i^*\left(t-\tau_{Bij}\right)e^{j2\pi\left(f_{Aij}-f_{Bij}\right)t}dt \qquad (1.43)$$

In general, we use the normalized $|\chi_{network}|^2$ to draw the fuzzy functional diagram and analyze the target resolution of the system. So, the fuzzy function of the system can be expressed as:

$$X_{network} = \frac{1}{\gamma}\left|\sum_{j=1}^{N}\sum_{i=1}^{M}\eta_{ij}\chi_{ij}\right|^2 \tag{1.44}$$

In this formula, γ is the normalization factor and χ_{ij} is the fuzzy function of the transceiver unit formed by the ith transmitting station and the jth receiving station.

If the passive reconnaissance information is considered, the fuzzy function of the system can be expressed as:

$$X_{H network} = \frac{1}{\gamma}\left|\sum_{j=1}^{N}\sum_{i=1}^{M}\eta_{ij}\chi_{ij} + \sum_{j=1}^{N}w_j\chi_{p_j}\right|^2 \tag{1.45}$$

In formula (1.45), χ_{P_j} is the fuzzy function of the transmitting signal of the target with emitter received by the jth receiving station, and w_j is the correlated coefficient, whose expression is:

$$w_j = \frac{P_{Rj}}{\text{Max}(P_{Rj})} \tag{1.46}$$

In formula (1.46), P_{Rj} is the power of the transmitting signal of the target with emitter received by the jth receiving station, whose expression is:

$$P_{Rj} = \frac{P_t'G_t'G_{rj}\lambda_p^2}{(4\pi R_{rj})^2} \tag{1.47}$$

In formula (1.47), P_t' is the transmitting power of the target's emitter, G_t' is the antenna gain of the target's emitter, λ_p is the wavelength of the transmitting signal of the target's emitter, and the other parameters are the same as those in formula (1.42).

1.4.6.2 Multiple Transmitter One Receiver Mode

When the system works in the multiple transmitter one receiver mode, i.e., several transmitting stations and one receiving station are working simultaneously, the receiving station can receive the signals of M transmitting stations and, at the same time, receive signals emitted by the target's emitters.

Fig. 1.36 Structure sketch diagram of two transmitters and one receiver in a network radar countermeasure system

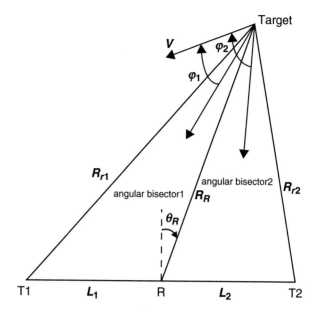

Fig. 1.37 $N = 5$ uniform pulse train signals

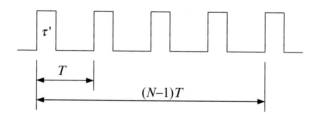

Now, let us take the deployment of two transmitting stations and one receiving station (two transmitters one receiver) as an example to illustrate the characteristics of fuzzy function in the multiple transmitter one receiver mode (Fig. 1.36)

Suppose the complex envelopes of transmitting signals of the two transmitting stations are of uniform pulse train, as shown in Fig. 1.37.

A single pulse can be expressed as:

$$s_d(t) = \frac{1}{\sqrt{\tau'}} Rect\left(\frac{t}{\tau'}\right) \tag{1.48}$$

Then, the uniform pulse train can be expressed as:

$$s(t) = \frac{1}{\sqrt{N}} \sum_{i=0}^{N-1} s_d(t - iT) \tag{1.49}$$

From the basic definition of fuzzy function, we can obtain:

$$
\begin{aligned}
\chi(\tau_A, \tau_B, f_A, f_B) &= \int_{-\infty}^{\infty} s(t - \tau_A) s^*(t - \tau_B) e^{j2\pi(f_A - f_B)t} dt \\
&= \frac{1}{N} \sum_{i=0}^{N-1} \sum_{j=0}^{N-1} \int_{-\infty}^{\infty} s_d(t - iT - \tau_A) s_d^*(t - jT - \tau_B) e^{j2\pi(f_A - f_B)t} dt
\end{aligned}
$$

$$(1.50)$$

By using t_1 to substitute $t - iT$, we get:

$$
\begin{aligned}
\chi(\tau_A, \tau_B, f_A, f_B) = \ &\frac{1}{N} \sum_{i=0}^{N-1} e^{j2\pi(f_A - f_B)iT} \sum_{j=0}^{N-1} \int_{-\infty}^{\infty} s_d(t_1 - \tau_A) s_d^*(t_1 - [\tau_B - (i - j)T]) \\
&\times e^{j2\pi(f_A - f_B)t_1} dt_1
\end{aligned}
$$

$$(1.51)$$

where:

$$
\begin{aligned}
\int_{-\infty}^{\infty} s_d(t_1 - \tau_A) s_d^*(t_1 - [\tau_B - (i - j)T]) e^{j2\pi(f_A - f_B)t_1} dt_1 \\
= \chi_d(\tau_A, \tau_B - (i - j)T, f_A, f_B)
\end{aligned}
$$

$$(1.52)$$

Then, formula (1.51) can be abbreviated as:

$$
\chi(\tau_A, \tau_B, f_A, f_B) = \frac{1}{N} \sum_{i=0}^{N-1} e^{j2\pi(f_A - f_B)iT} \sum_{j=0}^{N-1} \chi_d
$$

$$(1.53)$$

Let $q = i - j$, then:

$$
\sum_{i=0}^{N-1} \sum_{j=0}^{N-1} = \sum_{q=-(N-1)}^{0} \sum_{i=0}^{N-1-|q|} \Bigg|_{for\ j=i-q} + \sum_{q=1}^{N-1} \sum_{j=0}^{N-1-|q|} \Bigg|_{for\ i=j+q}
$$

$$(1.54)$$

Substituting (1.54) into (1.53), we can obtain:

$$
\begin{aligned}
\chi(\tau_A, \tau_B, f_A, f_B) = \ &\frac{1}{N} \sum_{q=-(N-1)}^{0} \left\{ \chi_d \sum_{i=0}^{N-1-|q|} e^{j2\pi(f_A - f_B)iT} \right\} \\
&+ \frac{1}{N} \sum_{q=1}^{N-1} \left\{ e^{j2\pi(f_A - f_B)qT} \chi_d \sum_{j=0}^{N-1-|q|} e^{j2\pi(f_A - f_B)jT} \right\}
\end{aligned}
$$

$$(1.55)$$

Let:

$$z = \exp(j2\pi(f_A - f_B)T) \tag{1.56}$$

and:

$$\sum_{j=0}^{N-1-|q|} z^j = \frac{1 - z^{N-|q|}}{1 - z} \tag{1.57}$$

then from formulas (1.55) and (1.57), we can obtain:

$$\sum_{i=0}^{N-1-|q|} e^{j2\pi(f_A-f_B)iT} = e^{j\pi(f_A-f_B)(N-1-|q|)T} \frac{\sin\left[\pi(f_A-f_B)(N-|q|)T\right]}{\sin\left[\pi(f_A-f_B)T\right]} \tag{1.58}$$

Substituting formula (1.58) into formula (1.55), we can obtain:

$$\chi(\tau_A, \tau_B, f_A, f_B) = \frac{1}{N} \sum_{q=-(N-1)}^{N-1} e^{[j\pi(f_A-f_B)(N-1+q)T]} \frac{\sin\left[\pi(f_A-f_B)(N-|q|)T\right]}{\sin\left[\pi(f_A-f_B)T\right]} \chi_d \tag{1.59}$$

In formula (1.59), χ_d is the fuzzy function of a single pulse.

Suppose that the complex envelope of the signal emitted by the target's emitter reconnoitered by the receiving station of the system is:

$$u(t) = \begin{cases} 1/\sqrt{T'} & (0 < t < T') \\ 0 & \text{(other conditions)} \end{cases}$$

then:

$$\chi(\tau_A, \tau_B, f_A, f_B) = \begin{cases} e^{j\pi f_d(T'-\tau)} \left[\dfrac{\sin\left(\pi f_d(T'-|\tau|)\right)}{\pi f_d(T'-|\tau|)}\right] \dfrac{(T'-|\tau|)}{T'}, & |\tau| < T' \\ 0, & |\tau| \geq T' \end{cases} \tag{1.60}$$

From formulas (1.59) and (1.60), we obtain:

$$\chi_d = \begin{cases} e^{j\pi f_d(\tau'-\tau)} \left[\dfrac{\sin\left(\pi f_d(\tau'-|\tau|)\right)}{\pi f_d(\tau'-|\tau|)}\right] \dfrac{(\tau'-|\tau|)}{\tau'}, & |\tau| < \tau' \\ 0, & |\tau| \geq \tau' \end{cases} \tag{1.61}$$

where $\tau = \tau_A - \tau_B + qT$ and $f_d = f_A - f_B$.

Assume that the target's echo signal received by the receiving station is a uniform pulse train $N=3$, with pulse width $\tau' = 40\,\mu s$, pulse repetition interval

Fig. 1.38 The geometric relationship between the target, the receiving station, and the transmitting station

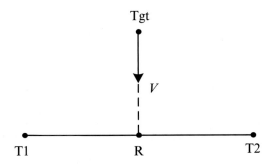

$T = 100\,\mu s$, and wavelength $\lambda = 6.28\,m$. The receiving station's visual angle of the target $\theta_R = 0^0$, the target velocity $V = 600\,m/s$, the direction is the connecting line between the target and the receiving station, which moves towards the direction of the receiving station, $R_R = 60\,km$, and $L_1 = L_2 = 100\,km$. Their geometric relation is as shown in Fig. 1.38.

For the convenience of easy discussion, we introduce the time delay and Doppler frequency shift of the transceiver unit while transmitting and receiving separately. The transmitting–receiving time delay can be expressed by R_R, θ_R, and L as:

$$\tau(R_R, \theta_R, L) = \left[R_R + \sqrt{R_R^2 + L^2 + 2R_RL\sin\theta_R}\right]/c \qquad (1.62)$$

Similarly, the transmitting–receiving time delay can also be expressed by R_R, θ_R, and L as:

$$\tau(R_T, \theta_T, L) = \left[R_T + \sqrt{R_T^2 + L^2 - 2R_TL\sin\theta_T}\right]/c \qquad (1.63)$$

When R_R and θ_R are known, the Doppler frequency shift of the transceiver unit f_D can be expressed as:

$$f_D(R_R, V\cos\phi, \theta_R, L) = 2\frac{f_c}{c}V\cos\phi\sqrt{\frac{1}{2} + \frac{R_R + L\sin\theta_R}{2\sqrt{R_R^2 + L^2 + 2R_RL\sin\theta_R}}} \qquad (1.64)$$

Similarly, when R_T and θ_T are known, the Doppler frequency shift of the transceiver unit f_D can be expressed as:

$$f_D(R_T, V\cos\phi, \theta_T, L) = 2\frac{f_c}{c}V\cos\phi\sqrt{\frac{1}{2} + \frac{R_T - L\sin\theta_T}{2\sqrt{R_T^2 + L^2 - 2R_TL\sin\theta_T}}} \qquad (1.65)$$

From formulas (1.62) and (1.64), we can obtain τ_A, τ_B, f_A, and f_B. By substituting them into formula (1.59), we can obtain the fuzzy function of the system's

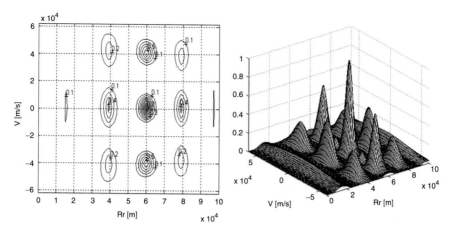

Fig. 1.39 Fuzzy function graph of the system in two transmitter one receiver mode

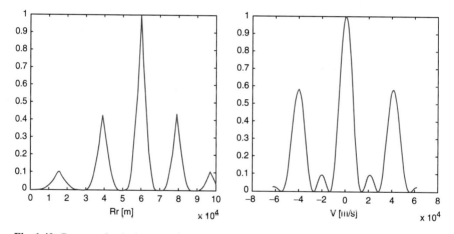

Fig. 1.40 Range and velocity resolution of the system in two transmitter one receiver mode

transceiver unit, and then from formula (1.44), we can obtain the fuzzy function of the system in the two transmitter one receiver mode, as shown in Fig. 1.39. (For the time being, we do not consider the passive reconnaissance information that the receiving station received χ_P.) The range and velocity resolution of the system in the two transmitter one receiver mode is shown in Fig. 1.40.

If we consider that the two transmitting stations transmit different signals, the signal parameter of transmitting station T_1 remains unchanged, while the transmitting signal of transmitting station T_2 is changed, with its the pulse repetition interval being $T = 126\,\mu s$ and the other parameters remaining unchanged. Now, we can obtain the fuzzy function diagram of the system, as is shown in Fig. 1.41. The range and velocity resolution in the two transmitter one receiver mode is shown in Fig. 1.42.

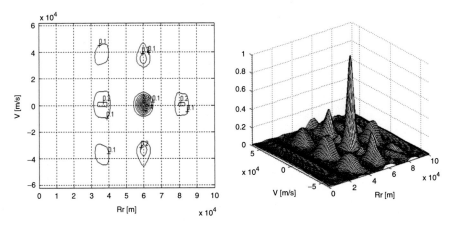

Fig. 1.41 Fuzzy function graph in two transmitter one receiver mode (different transmitting signals)

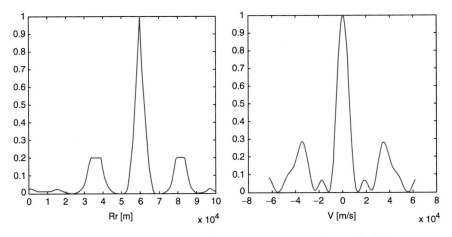

Fig. 1.42 Range and velocity resolution in two transmitter one receiver mode (different transmitting signals)

Comparing Figs. 1.42 and 1.41 with Figs. 1.40 and 1.39, we find that, when the transmitting signals of the two transmitting stations are different, the side lobes of the fuzzy function reduced significantly, which, to some extent, eliminated the periodic fuzziness of the range and velocity, thus improving the target resolution of the system.

If we consider the information of the target emitter received by the receiving station, we can obtain the fuzzy function of the system by using formula (1.45). Let the pulse width of the signal received by the receiving station of the target emitter be $\tau' = 20\,\mu s$, the pulse repetition interval $T = 125\,\mu s$, and the wavelength $\lambda = 3.14\,m$.

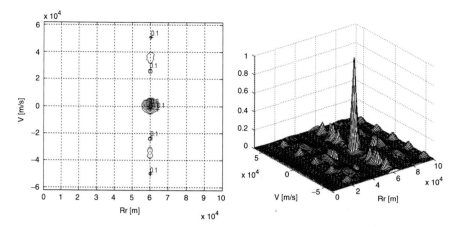

Fig. 1.43 Fuzzy function graph in two transmitter one receiver mode (different transmitting signals, using passive reconnaissance information)

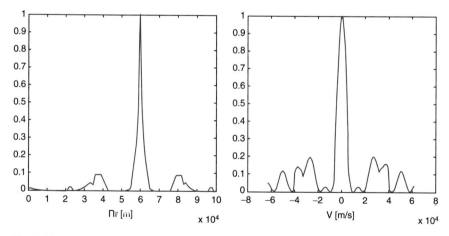

Fig. 1.44 Range and velocity resolution in two transmitter one receiver mode (different transmitting signals, using passive reconnaissance information)

We obtain the fuzzy functional graph in the two transmitter one receiver mode with different transmitting signals and passive reconnaissance information, as shown in Fig. 1.43; we also obtain the range and velocity resolution in the same mode, as shown in Fig. 1.44.

From Figs. 1.44 and 1.43, we can see that, after the use of passive reconnaissance information, the side lobe of the fuzzy function is further reduced and the peak narrowed, but the range and velocity resolution increased, showing that the passive reconnaissance information obtained at this time has improved the target resolution of the system.

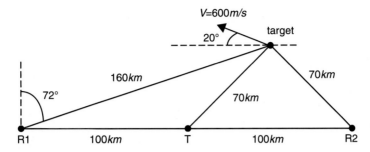

Fig. 1.45 Structure of the network radar countermeasure system in one transmitter two receiver mode

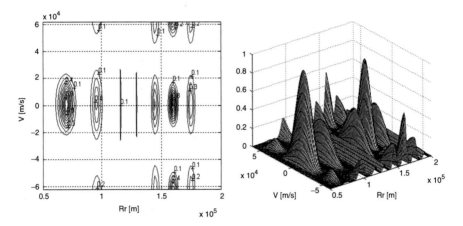

Fig. 1.46 Fuzzy function graph in one transmitter two receiver mode

1.4.6.3 One Transmitter Multiple Receiver Mode

When the system works in the one transmitter multiple receiver mode, i.e., one transmitting station and several receiving stations working simultaneously, N receiving stations can receive signals from the one and the same transmitting station and, at the same time, each receiving station can receive the signals from the target with emitters.

Take the configuration of one transmitting station and two receiving stations (one transmitter two receivers) as an example to illustrate the characteristics of the fuzzy function in one transmitter multiple receiver mode (Fig. 1.45).

Assume that the echo signals of the target received by the receiving station are the same as those in the simulation in Fig. 1.39, and the other parameters are as shown in Fig. 1.46. In the same way, from formulas (1.61) and (1.63), we can get τ_A, τ_B, f_A, and f_B. By substituting them into formulas (1.60) and (1.44), we can obtain the fuzzy function of the network radar countermeasure system in the one transmitter two receiver mode. Now, let us consider a simpler case. Let the weighted

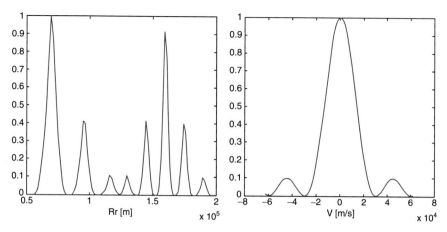

Fig. 1.47 Range and velocity resolution in one transmitter two receiver mode

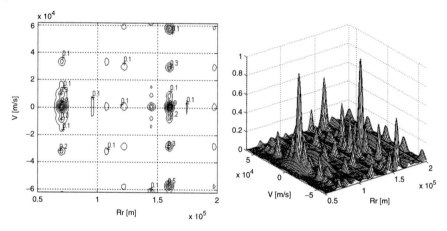

Fig. 1.48 Fuzzy function graph in one transmitter two receiver mode (using passive reconnaissance information)

coefficients η_{ij} and w_j be 1, without considering the passive reconnaissance information received by the receiving station. At this time, the fuzzy function is shown in Fig. 1.46, and the range and velocity resolution in the one transmitter two receiver mode is shown in Fig. 1.47.

If we consider the passive reconnaissance information obtained by the receiving station, from formula (1.45) we can obtain the fuzzy function graph, as shown in Fig. 1.48. If the target emitter's signals received by the receiving station are the same as above, then the range and velocity resolution in the one transmitter two receiver mode is shown in Fig. 1.49.

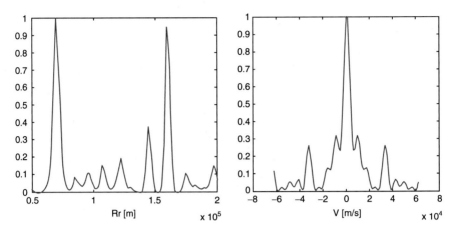

Fig. 1.49 Range and velocity resolution in one transmitter two receiver mode (using passive reconnaissance information)

From the results of the simulation, we can see that the passive reconnaissance information obtained can improve the target resolution of the system.

In addition, from the fuzzy function graphs in Figs. 1.46 and 1.48, we can see that, when the system works in the one transmitter multiple receiver mode, the fuzzy function is the sum of the fuzzy functions of the multiple transceiver units. In these figures, the two main peaks of the fuzzy function correspond to the peak of the fuzzy function of the two transceiver units. This indicates that the fuzzy function in the one transmitter multiple receiver mode reflects the target resolution of each transceiver unit. Therefore, we can select a specific transceiver unit to solve the fuzzy problem, but choose another transceiver unit to provide sufficient resolution of the target.

By substituting τ_A, τ_B, f_A, and f_B obtained in formulas (1.62) and (1.64) into formula (1.44), we can also obtain the fuzzy function of the system, which is the function of the target velocity and the range between the target and the transmitting station R_t. Similarly, if we do not consider the passive reconnaissance information received by the receiving station, the fuzzy function in the one transmitter two receiver mode is shown in Fig. 1.50, and the range and velocity resolution at the transmitting station in the one transmitter two receiver mode is shown in Fig. 1.51.

In Figs. 1.50 and 1.51, the fuzzy function as well as the range and velocity resolution have only one main peak, which indicates that the fuzzy function based on the transmitting station reflects the target resolution in one transmitter multiple receiver mode.

By reducing the baseline length of the two transceiver units in Fig. 1.45, we can obtain the fuzzy function and the range and velocity resolution of the system by simulation. Let the baseline length of the two transceiver units be $L = 10$ km and the other parameters remain unchanged.

When the length of the baseline $L = 10$ km, the fuzzy function graph at the transmitting station in the one transmitter two receiver mode is as shown in

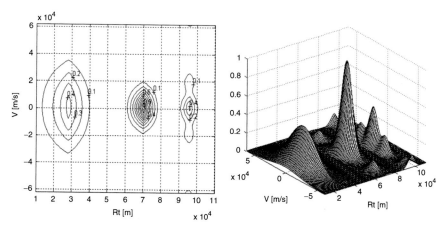

Fig. 1.50 Fuzzy function at the transmitting station in one transmitter two receiver mode

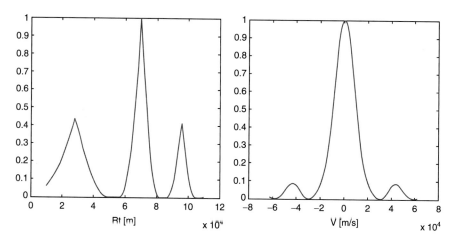

Fig. 1.51 Range and velocity resolution at the transmitting station in one transmitter two receiver mode

Fig. 1.52, and the range and velocity resolution at the transmitting station in the one transmitter two receiver mode is as shown in Fig. 1.53.

Comparing Figs. 1.53 and 1.52 with Figs. 1.51 and 1.50, we can see that reducing the baseline length of the transceiver units makes the main peak of the fuzzy function narrower, the range and velocity resolution increased, and the target resolution improved.

If we consider the passive reconnaissance information obtained by the receiving station, plus the use of formula (1.45), we can obtain the fuzzy function graph of the

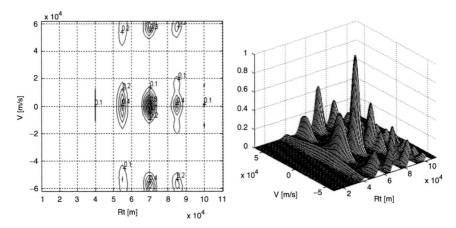

Fig. 1.52 Fuzzy function at the transmitting station in one transmitter two receiver mode ($L = 10$ km)

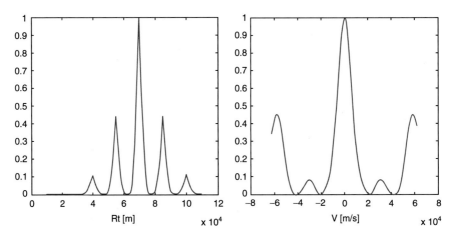

Fig. 1.53 Range and velocity resolution at the transmitting station in one transmitter two receiver mode ($L = 10$ km)

system, as shown in Fig. 1.54 and the range and velocity resolution based on the transmitting station as shown in Fig. 1.55.

From Figs. 1.54 and 1.55, we can see that the use of the obtained passive reconnaissance information has improved the target resolution.

In addition, Figs. 1.50, 1.54, and 1.52 have shown that the fuzzy function is based on the transmitting station, which explains the system's target resolution on the whole. However, Figs. 1.46 and 1.48 have expressed that the fuzzy function explains the system's target resolution from the angle of every locality, i.e., every

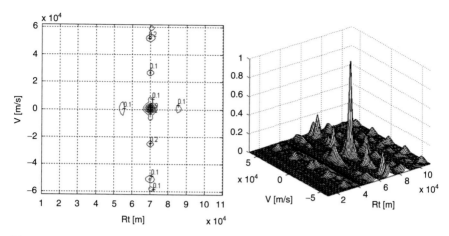

Fig. 1.54 Fuzzy function at the transmitting station in one transmitter two receiver mode (using passive reconnaissance information)

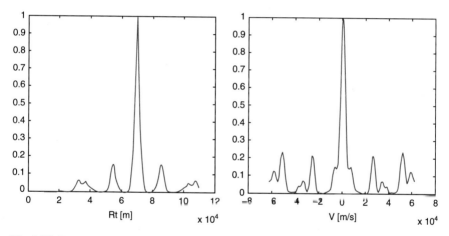

Fig. 1.55 Range and velocity resolution at the transmitting station in one transmitter two receiver mode (using passive reconnaissance information)

transceiver. Consequently, we can select different fuzzy functional representations of the system according to the actual needs.

1.4.6.4   Multiple Transmitters Multiple Receivers Mode

The system is typically operated in the multiple transmitters multiple receivers mode, i.e., M transmitting stations and N receiving stations are working simultaneously, and every receiving station can not only receive the signals transmitted

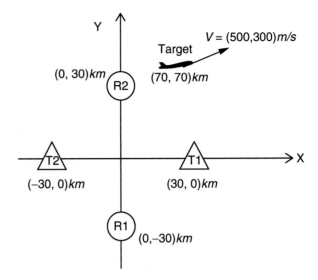

Fig. 1.56 Structure of the network radar countermeasure system in two transmitter two receiver mode

by the M transmitting stations, but also the signals emitted by the target with emitters.

Take the configuration of two transmitting stations and two receiving stations (two transmitters, two receivers) as an example to illustrate the characteristics of its fuzzy function in the multiple transmitters multiple receivers mode.

In Fig. 1.56, T represents the transmitting station, R represents the receiving station, and V represents the target speed. The diagram shows the position coordinates of the transmitting station, the receiving station, and the target, but also gives the components of the target speed in the X-axis and Y-axis.

The time delay τ and Doppler frequency shift f_D of the echo signal can be represented by the transmitting station, the receiving station, the position coordinates of the target, and the velocity of the target, namely:

$$\tau(x, y) = (R_T + R_R)/c$$
$$= \left[\sqrt{(tx_k - x)^2 + (ty_k - y)^2} + \sqrt{(rx_l - x)^2 + (ry_l - y)^2}\right]/c \quad (1.66)$$

$$f_D(x, y, v_x, v_y) = \left[\frac{(tx_k - x)v_x + (ty_k - y)v_y}{R_T} + \frac{(rx_l - x)v_x + (ry_l - y)v_y}{R_R}\right]\frac{1}{\lambda} \quad (1.67)$$

In the above formulas, (tx_k, ty_k) is the coordinate of the kth transmitting station, (rx_l, ry_l) is the coordinate of the lth receiving station, (x, y) is the position of the target, (v_x, v_y) is the component of the target velocity in the X and Y directions, λ is the wavelength of the signal, and c is the speed of light.

By substituting formulas (1.62) and (1.63) into the expressions of the fuzzy function, we can obtain the relationship between the fuzzy function of the network radar countermeasure system, the target position (x, y), and the target velocity $(v_x,$

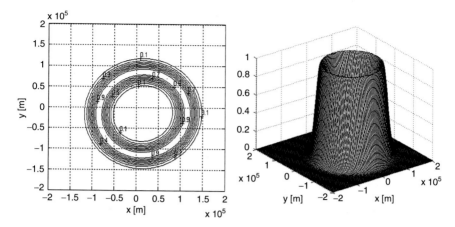

Fig. 1.57 The range fuzzy function diagram of the transceiver unit composed of T_1 and R_1

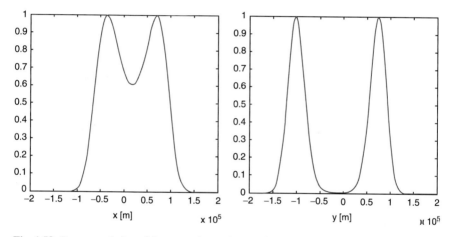

Fig. 1.58 Range resolution of the transceiver unit composed of T_1 and R_1

v_y). To visualize the overall fuzzy function chart in the two transmitter two receiver mode, we have drawn separately the distance fuzzy function chart (the relationship between the fuzzy function and the target position) and the velocity fuzzy function chart (the relationship between the fuzzy function and the target speed), in order to study separately the distance resolution and velocity resolution.

Assuming that the signals transmitted by the two transmitting stations are the same, the wavelength is $\lambda = 6.28$ m and $T = 1 \times 10^{-4}$ s. The other parameters are as shown in Fig. 1.57, and the range fuzzy function and range resolution of the transceiver unit consisting of T_1 and R_1 are shown in Figs. 1.58 and 1.59 (without considering the obtained passive reconnaissance information for now).

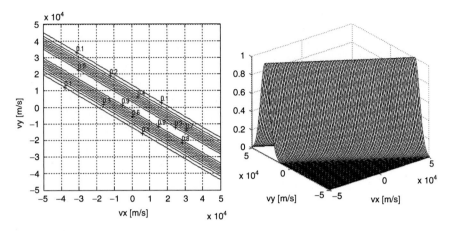

Fig. 1.59 Velocity fuzzy function of the transceiver unit composed of T_1 and R_1

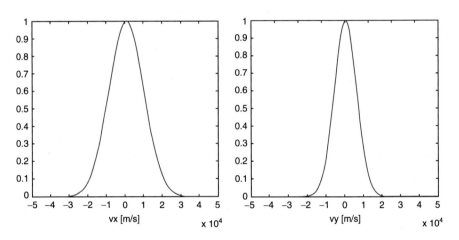

Fig. 1.60 Velocity resolution of the transceiver unit composed of T_1 and R_1

The velocity fuzzy function and velocity resolution of the transceiver composed of T_1 and R_1 are shown in Figs. 1.59 and 1.60.

The range fuzzy function and range resolution in the two transmitter two receiver mode (Fig. 1.56) are shown in Figs. 1.61 and 1.62.

The velocity fuzzy function and velocity resolution in the two transmitter two receiver mode are shown in Figs. 1.63 and 1.64.

Comparing Figs. 1.58 and 1.60 with Figs. 1.62 and 1.64, we can see that the range resolution and velocity resolution in the two transmitter two receiver mode are superior to those of the transceiver unit, indicating that adding transmitting and receiving stations can improve the target resolution of the system.

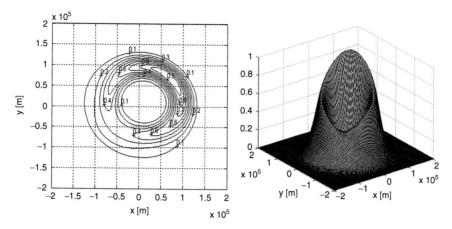

Fig. 1.61 Range fuzzy function in two receiver two transmitter mode

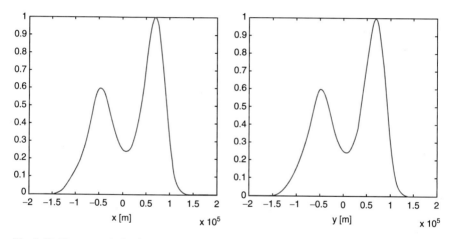

Fig. 1.62 Range resolution in two receiver two transmitter mode

The velocity fuzzy function and velocity resolution of the transceiver unit composed of T_1 and R_1 are shown in Figs. 1.59 and 1.60.

The range fuzzy function and range resolution in the two receiver two transmitter mode are shown in Figs. 1.61 and 1.62.

The velocity fuzzy function and velocity resolution in the two receiver two transmitter mode are shown in Figs. 1.63 and 1.64.

From the comparison of Figs. 1.58, 1.60, 1.62, and 1.64, we can see that the range resolution and the velocity resolution in the two receiver two transmitter mode are both better than those of the transceiver unit composed of T_1 and R_1, which shows that increasing the number of transmitters and receivers could help improve the system's resolution.

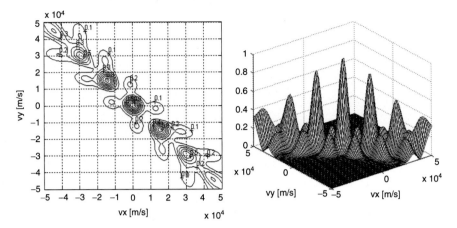

Fig. 1.63 Velocity fuzzy function in two receiver two transmitter mode

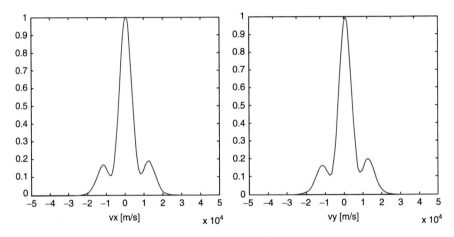

Fig. 1.64 Velocity resolution in two receiver two transmitter mode

Next, we consider what the receiving station receives is the passive reconnaissance information, and use formula (1.45) to obtain the range fuzzy function and resolution in the two transmitter two receiver mode as shown in Figs. 1.65 and 1.66. Suppose that the signal pattern of the target emitter received by the receiving station is the same as that transmitted by the transmitting station, but $T = 0.1 \times 10^{-4}$ s, and the other parameters remain unchanged.

At this time, the velocity fuzzy function and resolution are shown in Figs. 1.67 and 1.68, respectively.

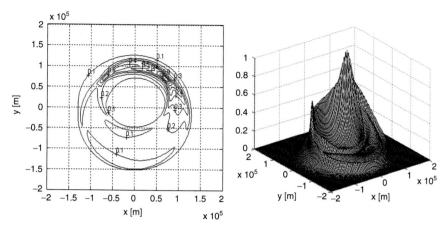

Fig. 1.65 Range fuzzy function in two receiver two transmitter mode (using passive reconnaissance information)

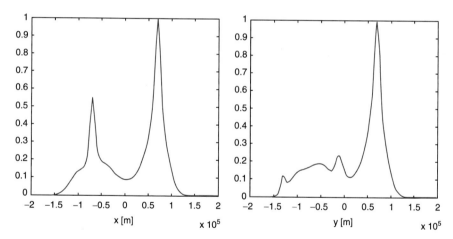

Fig. 1.66 Range resolution in two receiver two transmitter mode (using passive reconnaissance information)

From Figs. 1.65, 1.66, 1.67, and 1.68, we can see that the peak of the fuzzy function was further narrowed and the range and velocity resolution are further improved, also showing that the passive reconnaissance information received at this time had improved the target resolution of the system.

From the above simulation analysis, we can reach a conclusion that the fuzzy function and target resolution of the network radar countermeasure system are not only related to the waveform of the transmitting signal of the transmitting station, but are also closely related to other factors, such as the structure of the system, the baseline length of the transceiver, the relative position of the target, the baseline of

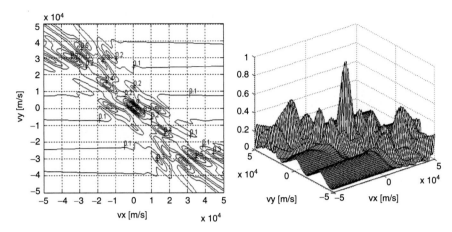

Fig. 1.67 Velocity fuzzy function in two receiver two transmitter mode (using passive reconnaissance information)

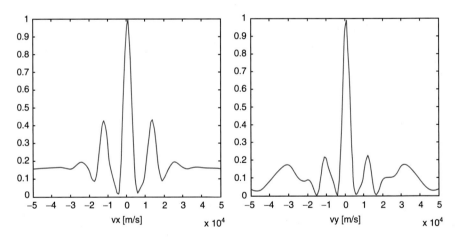

Fig. 1.68 Velocity resolution in two receiver two transmitter mode (using passive reconnaissance information)

the transceiver, the number of transmitting and receiving stations, as well as the obtained passive reconnaissance information.

Chapter 2
Target Positioning of Network Radar Countermeasure Systems

2.1 Introduction

The network radar countermeasure system can be divided into the following: active detection mode, passive reconnaissance mode, and active passive integrated mode. In each mode, the principle and process of target location are different. When working in an active detection mode, the system will use the target to detect and locate the echo signal of transmitting signals and the receiving station can obtain the target state quantity measurement information, including target arrival time (TOA, or range and angle), information (azimuth and elevation), Doppler, etc. At this point, the system has strong abilities in target positioning; when working in a passive detection mode, the system can locate the target by intercepting the target radiation signal. The moving state quantity measuring information intercepted by the receiving station includes the measuring pulse arrival time (TOA), the angle information (azimuth and elevation), and target attribute quantity measurement information (the radiation source frequency, pulse width, antenna scanning frequency, pulse cycle, pulse characteristic information, etc.). The positioning accuracy depends on the system configuration and the accuracy of the time difference.

There is one point that we should make clear. That is, in the active mode and passive mode, positioning is available to guide one other, to increase intercept probability and improve the target data rate. The integration mode is conducive to the realization of target data fusion and target recognition.

© National Defense Industry Press, Beijing and Springer-Verlag Berlin Heidelberg 2016   67
Q. Jiang, *Network Radar Countermeasure Systems*,
DOI 10.1007/978-3-662-48471-5_2

2.2 Active Mode Target Location

The active mode can be divided into the one transmitter multiple receiver mode, multiple transmitter multiple receiver mode, and multiple transmitter one receiver mode. The location processing of the one transmitter multiple receiver and multiple transmitter multiple receiver modes includes the two structures of the distributed centralized mode and centralized mode.

1. Distributed centralized structure: The system is regarded as one consisting of m $(m > 1)$ receiving stations (common $n(n \geq 1)$ transmitting stations). Each receiving station first conducts positioning processing independently, and then the results are transmitted to the network center station, where the position fusion processing is conducted. In other words, the structure carries out further fusion processing of the positioning results of the one transmitter multiple receiver mode and the multiple transmitter multiple receiver mode in the network center station so as to obtain better quality of the target position estimation. If needed, the information after the fusion process can be fed back to each network radar countermeasure system receiving station for the system calibration process.
2. Centralized structure: The measurement data from each receiving station are transmitted directly to the network center station, where the data are centralized for target location processing.

2.2.1 Multiple Transmitter One Receiver Mode

2.2.1.1 Acquisition of the Positioning Solution

Suppose if the transmitting station is $T_i(x_{T_i}, y_{T_i}, z_{T_i}), i = 1, 2 \cdots, n(n \geq 1)$, the receiving station is $R_l(x_R, y_R, z_R), l = 1, 2 \cdots, m$, and then the measurement of one receiving station is as follows:

$$
\begin{cases}
r_{T_i} = \sqrt{(x - x_{T_i})^2 + (y + y_{T_i})^2 + (z - z_{T_i})^2} (i = 1, 2, \ldots, n) \\
r_{R_i} = \sqrt{(x - x_{R_i})^2 + (y - y_{R_i})^2 + (z - z_{R_i})^2} \\
\rho_{R_i T_i} = r_{R_i} + r_T, (i = 1, 2, \cdots, n(n \geq 2); l = 1 \text{ or } 2, \cdots, \text{ or } m) \\
\varphi_{R_i} = \text{arctg} \dfrac{y - y_{R_i}}{x - x_{R_i}} \\
\varepsilon_{R_i} = \text{arctg} \dfrac{z - z_{R_i}}{\sqrt{(x - x_{R_i})^2 + (y - y_{R_i})^2}}
\end{cases}
\tag{2.1}
$$

Organizing the distance and the equation of formula (2.1), we can get:

$$(x_{T_i} - x_{R_l})x + (y_{T_i} - y_{R_l})y + (z_{T_i} - z_{R_l})z = k_{R_l} - \rho_{R_lT_i}(\rho_{R_lT_i} - r_{R_l}) \tag{2.2}$$

Among that, $k_{R_l} \overset{\Delta}{=} \frac{1}{2}\left[\rho_{R_lT_i}^2 + \left(x_{T_i}^2 + y_{T_i}^2 + z_{T_i}^2\right) - \left(x_{R_l}^2 + y_{R_l}^2 + z_{R_l}^2\right)\right](i = 1, 2, \cdots, n)$

Organizing the equation for the azimuth angle measurement, we can obtain:

$$x\ \sin\phi_{R_l} - y\ \cos\phi_{R_l} = x_{R_l}\sin\phi_{R_l} - y_{R_l}\cos\phi_{R_l} \tag{2.3}$$

The measurement equation of the pitch angle can be obtained as follows:

$$\begin{aligned} y\sin\varepsilon_{R_l} - z\sin\varphi_{R_l}\cos\varepsilon_{R_l} &= y_{R_l}\sin\varepsilon_{R_l} - z_{R_l}\sin\varphi_{R_l}\cos\varepsilon_{R_l}\ \text{or} \\ x\sin\varepsilon_{R_l} - z\cos\varphi_{R_l}\cos\varepsilon_{R_l} &= x_{R_l}\sin\varepsilon_{R_l} - z_{R_l}\cos\varphi_{R_l}\cos\varepsilon_{R_l}\ 0 \end{aligned} \tag{2.4}$$

The formulas (2.2), (2.3), and (2.4) constitute a non-linear equation, where r_{R_l} is the non-linear function of x, y, z; therefore, the display of the x, y, z shows that expression is difficult. Here, the indirect method is adopted, i.e., first, r_{R_l} is regarded as a known quantity, thus the solution in terms of x, y, z is the linear function of r_{R_l}, and then substitute x, y, z with the r_{R_l} expressions to obtain r_{R_l}, and then make the solution traverse through the azimuth parameter to obtain the ambiguity. The concrete process is as follows:

A matrix form of formulas (2.2), (2.3), and (2.4) is:

$$AX = F \tag{2.5}$$

In the formula:

$$A = \begin{bmatrix} x_{T_1} - x_{R_l} & y_{T_1} - y_{R_l} & z_{T_1} - z_{R_l} \\ \vdots & \vdots & \vdots \\ x_{T_n} - x_{R_l} & y_{T_n} - y_{R_l} & z_{T_n} - z_{R_l} \\ \sin\phi_{R_l} & -\cos\phi_{R_l} & 0 \\ 0 & \sin\varepsilon_{R_l} & -\sin\phi_{R_l}\cos\varepsilon_{R_l} \end{bmatrix}_{(n+2)\times 3}$$

$$X = [x, y, z]^T$$

$$F = \begin{bmatrix} k_{R_l} - \rho_{R_lT_1}(\rho_{R_lT_1} - r_{R_l}) \\ \vdots \\ k_{R_l} - \rho_{R_lT_n}(\rho_{R_lT_n} - r_{R_l}) \\ x_{R_l}\sin\varphi_{R_l} - y_{R_l}\cos\varphi_{R_l} \\ y_{R_l}\sin\varepsilon_{R_l} - z_{R_l}\sin\varphi_{R_l}\cos\varepsilon_{R_l} \end{bmatrix}_{(n+2)\times 1}$$

Under the conditions of suitable site selection, $\text{rank}(A) = 3$, and using the pseudo inverse method to solve (2.5), we can obtain:

$$\hat{X} = \left(A^T A\right)^{-1} A^T F \tag{2.6}$$

Suppose that:

$$\left(A^T A\right)^{-1} A^T \triangleq \begin{bmatrix} a_{11} & \cdots & a_{1(n+2)} \\ a_{21} & \cdots & a_{2(n+2)} \\ a_{31} & \cdots & a_{3(n+2)} \end{bmatrix} = \left[a_{ij}\right]_{3\times(n+2)} \tag{2.7}$$

From formula (2.6), we can obtain:

$$
\begin{cases}
\hat{x} = \displaystyle\sum_{j=1}^{n} a_{1j} k_{R_l} - \sum_{j=1}^{n} \rho_{R_l T_j}^2 + a_{1(n+1)}\left(x_{R_l}\ \sin\varphi_{R_l} - y_{R_l}\ \cos\varphi_{R_l}\right) \\[2mm]
\qquad + a_{1(n+2)}\left(y_{R_l}\ \sin\varepsilon_{R_l} - z_{R_l}\ \sin\varphi_{R_l}\ \cos\varepsilon_{R_l}\right) + \left(\displaystyle\sum_{j=1}^{n} a_{1j}\rho_{R_l T_j}\right) r_{R_l} \overset{\Delta}{=} m_1 + n_1 r_{R_l} \\[4mm]
\hat{y} = \displaystyle\sum_{j=1}^{n} a_{2j} k_{R_l} - \sum_{j=1}^{n} \rho_{R_l T_j}^2 + a_{2(n+1)}\left(x_{R_l}\ \sin\varphi_{R_l} - y_{R_l}\ \cos\varphi_{R_l}\right) \\[2mm]
\qquad + a_{2(n+2)}\left(y_{R_l}\ \sin\varepsilon_{R_l} - z_{R_l}\ \sin\varphi_{R_l}\ \cos\varepsilon_{R_l}\right) + \left(\displaystyle\sum_{j=1}^{n} a_{2j}\rho_{R_l T_j}\right) r_{R_l} \overset{\Delta}{=} m_2 + n_2 r_{R_l} \\[4mm]
\hat{z} = \displaystyle\sum_{j=1}^{n} a_{3j} k_{R_l} - \sum_{j=1}^{n} \rho_{R_l T_j}^2 + a_{3(n+1)}\left(x_{R_l}\ \sin\varphi_{R_l} - y_{R_l}\ \cos\varphi_{R_l}\right) \\[2mm]
\qquad + a_{3(n+2)}\left(y_{R_l}\ \sin\varepsilon_{R_l} - z_{R_l}\ \sin\varphi_{R_l}\ \cos\varepsilon_{R_l}\right) + \left(\displaystyle\sum_{j=1}^{n} a_{3j}\rho_{R_l T_j}\right) r_{R_l} \overset{\Delta}{=} m_3 + n_3 r_{R_l}
\end{cases}
\tag{2.8}
$$

Among that:

$$
\begin{cases}
m_i = \displaystyle\sum_{j=1}^{n} a_{ij} k_{R_l} - \sum_{j=1}^{n} \rho_{R_l T_j}^2 + a_{i(n+1)}\left(x_{R_l}\ \sin\varphi_{R_l} - y_{R_l}\ \cos\varphi_{R_l}\right) \\[2mm]
\qquad + a_{i(n+2)}\left(y_{R_l}\ \sin\varepsilon_{R_l} - z_{R_l}\ \sin\varphi_{R_l}\ \cos\varepsilon_{R_l}\right) \qquad\qquad (i = 1,\ 2,\ 3) \\[3mm]
\qquad\qquad n_i = \displaystyle\sum_{j=1}^{n} a_{ij}\rho_{R_l T_j}
\end{cases}
\tag{2.9}
$$

Applying formula (2.7) into the r_{R_l1} expressions, we can obtain:

$$a r_{R_l}^2 - 2b r_{R_l} + c = 0 \tag{2.10}$$

In the above formula:

$$\begin{cases} a = n_1^2 + n_2^2 + n_3^2 - 1 \\ b = n_1(m_1 - x_{R_l}) + n_2(m_2 - y_{R_l}) + n_3(m_3 - z_{R_l}) \\ c = (m_1 - x_{R_l})^2 + (m_2 - y_{R_l})^2 + (m_3 - z_{R_l})^2 \end{cases} \quad (2.11)$$

The two values r_{R_l1} and r_{R_l2} of the r_{R_l1} can be solved by (2.9), and there exists the position ambiguity. If r_{R_l1} and r_{R_l2} are one negative and one positive, then only take the positive value. Otherwise, we can combine the azimuth parameter to solve for the ambiguity. The specific methods are as follows: r_{R_l1} and r_{R_l2} will be substituted with the formula (2.7), and we can estimate two points $(\hat{x}_1, \hat{y}_1, \hat{z}_1)$ and $(\hat{x}_2, \hat{y}_2, \hat{z}_2)$ of the target location. By calculation, we can obtain two azimuths $\hat{\theta}_1$ and $\hat{\theta}_2$ with respect to the receiving station. If $|\hat{\theta}_1 - \theta_{R_l}| \leq |\hat{\theta}_2 - \theta_{R_l}|$, we take $\hat{X} = [\hat{x}_1, \hat{y}_1, \hat{z}_1]^T$; otherwise, we take $\hat{X} = [\hat{x}_2, \hat{y}_2, \hat{z}_2]^T$, thereby eliminating the ambiguity in positioning.

2.2.1.2 Positioning Error Analysis

Assuming that the measurement errors are Gaussian white noise with zero mean, which are not correlated with each other. Their standard deviation corresponding to distance and azimuth angle, and the pitching angle measurement are $\delta_{\rho_{R_lT_1}}$, $\delta_{\varphi_{R_l}}$, and $\delta_{\varepsilon_{R_l}}$. Location error standard deviation is δ_s. Location errors between each component, each location error and each observation error are not correlated. During the process of analysis, the observation error is generally assumed to be the value maximum, so the standard deviation is constant.

By differentiating formula (2.1), we get:

$$\begin{aligned} d\rho_{R_lT_i} &= (c_{T_i1} + c_{R_l1})dx + (c_{T_i2} + c_{R_l2})dy \\ &\quad + (c_{T_i3} + c_{R_l3})dz + (k_{R_l} + k_{T_i})(i = 1, 2, \cdots, n) \\ d\phi_{R_l} &= -\frac{\sin^2\phi_{R_l}}{y - y_{R_l}}dx + \frac{\cos^2\phi_{R_l}}{x - x_{R_l}}dy + k_{\phi, R_l} \\ d\varepsilon_{R_l} &= -\frac{c_{R_l3}\cos\phi_{R_l}}{r_{R_l}}dx - \frac{c_{R_l3}\sin\phi_{R_l}}{r_{R_l}}dy + \frac{\cos\varepsilon_{R_l}}{r_{R_l}}dz + k_{\varepsilon, R_l} \end{aligned} \quad (2.12)$$

Among that:

$$c_{j1} = \frac{x - x_j}{r_j}, c_{j2} = \frac{y - y_j}{r_j}, c_{j3} = \frac{z - z_j}{r_j} \ (j = R_l, T_i) \quad (2.13)$$

$$\begin{cases} k_j = -\Big(c_{j1}dx_j + c_{j2}dy_j + c_{j3}dz_j\Big), \quad (j = R_l, T_i)(i = 1, 2, \cdots, n) \\[2mm] k_{\phi, R_l} = \dfrac{\sin^2\phi_{R_l}}{y - y_{R_l}}dx_{R_l} - \dfrac{\cos^2\phi_{R_l}}{x - x_{R_l}}dy_{R_l} \\[3mm] k_{\varepsilon, R_l} = \dfrac{c_{R_l3}\ \cos\phi_{R_l}}{r_{R_l}}dx_{R_l} + \dfrac{c_{R_l3}\ \sin\phi_{R_l}}{r_{R_l}}dy_{R_l} - \dfrac{\cos\varepsilon_{R_l}}{r_{R_l}}dz_{R_l} \end{cases} \tag{2.14}$$

The equation group composed of formula (2.11) is written in vector matrix form:

$$dY = CdX + dX_s \tag{2.15}$$

In the formula, we have:

$$\begin{cases} dY = \big[d\rho_{R_lT1}, \cdots, d\rho_{R_lT_n}, d\varphi_{R_i}, d\varepsilon_{R_i}\big]^T \text{---Observation error vector} \\[2mm] dX = [dx, dy, dz]^T \text{---Target position vector} \\[2mm] dX_s = \big[(k_{R_i} + k_{T_1}), \cdots, (k_{R_i} + k_{T_n}), k_{\varphi, R_i}, k_{\varepsilon, R_i}\big]^T \text{---The vector associated with site error} \end{cases}$$

$$C = \begin{bmatrix} c_{T_11} + c_{R_l1} & c_{T_12} + c_{R_l2} & c_{T_13} + c_{R_l3} \\ \vdots & \vdots & \vdots \\ c_{T_n1} + c_{R_l1} & c_{T_n2} + c_{R_l2} & c_{T_n3} + c_{R_l3} \\ -\dfrac{\sin^2\phi_{R_l}}{y - y_{R_l}} & \dfrac{\cos^2\phi_{R_l}}{x - x_{R_l}} & 0 \\ -\dfrac{c_{R_l3}\cos\phi_{R_l}}{r_{R_l}} & -\dfrac{c_{R_l3}\sin\phi_{R_l}}{r_{R_l}} & \dfrac{\cos\varepsilon_{R_l}}{r_{R_l}} \end{bmatrix}_{(n+2)\times 3} \tag{2.16}$$

C is the coefficient matrix related to the location of the site and the target.

The positioning error of the target is estimated by the pseudo inverse method via formula (2.15) as follows:

$$d\hat{X} = \big(C^TC\big)^{-1}C^T[dY - dX_s] \tag{2.17}$$

Suppose that:

$$\big(C^TC\big)^{-1}C^T \triangleq B \tag{2.18}$$

Because the R_l station itself has the timing error, together with the delay jitter error of target scattering, there exist some common errors in the distance measurement by the receiving station and the data, so the observation error of the $\rho_{R_lT_i}$ is related to each other. After setting the range and measurement error correction system as a zero mean system and leaving the site location error in each measurement unchanged, the location error components are uncorrelated with each other and each of the site location errors are uncorrelated, so the positioning error covariance is $\big(\because E[d\hat{X}] = 0\big)$

$$\big(dx_j, dy_j, dz_j\big)\ \big(j = T_i'\, \text{or}\, R_i\big)$$

$$P_{d\hat{X}} = E\left[d\hat{X}\,d\hat{X}^T\right]$$
$$= B\{E[dY\,dY^T] + E[dX_s\,dX_s^T]\}B^T \tag{2.19}$$

In this formula:

$$E[dY\,dY^T] = \begin{bmatrix} \delta^2_{\rho_{R_lT_1}} & \eta_{12}\delta_{\rho_{R_lT_1}}\delta_{\rho_{R_lT_2}} & \cdots & \eta_{1n}\delta_{\rho_{R_lT_1}}\delta_{\rho_{R_lT_n}} & 0 & 0 \\ & \delta^2_{\rho_{R_lT_2}} & \cdots & \eta_{2n}\delta_{\rho_{R_lT_2}}\delta_{\rho_{R_lT_n}} & 0 & 0 \\ - & & \ddots & & & \\ & & & \delta^2_{\rho_{R_lT_n}} & 0 & 0 \\ 0 & 0 & \cdots & 0 & \delta^2_{\varphi_{R_l}} & 0 \\ 0 & 0 & \cdots & 0 & 0 & \delta^2_{\varepsilon_{R_l}} \end{bmatrix} \tag{2.20}$$

where $\delta_{\rho_{R_lT_i}}$ is the standard deviation of the range and the measurement error of site R_l and η_{ij} is the correlation coefficient between the distance and the measurement error between the T_i, T_j site with the R_l site:

$$\eta_{ij} = \frac{\mathrm{cov}\left(d\rho_{R_iT_i}, d\rho_{R_jT_j}\right)}{\delta_{\rho_{R_iT_i}}\delta_{\rho_{R_jT_j}}} = \frac{E\left[d\rho_{R_iT_i}d\rho_{R_jT_j}\right]}{\delta_{\rho_{R_iT_i}}\delta_{\rho_{R_jT_j}}} \tag{2.21}$$

Assuming that the standard deviation of the site error components are the same, which is $\delta^2_{xj} = \delta^2_{yj} = \delta^2_{zj} = \delta^2_s$, and because $c^2_{j1} + c^2_{j2} + c^2_{j3} = 1$ ($j = T_i$ or R_i), can get:

$$E[dX_s\,dX_s^T] = \begin{bmatrix} & & & 0 & 0 \\ & \delta^2_s[I_n + q_n] & & \vdots & \vdots \\ & & & 0 & 0 \\ 0 & \cdots & 0 & \dfrac{\delta^2_s}{(x-x_{R_l})^2 + (y - y_{R_l})^2} & 0 \\ 0 & \cdots & 0 & 0 & \dfrac{\delta^2_s}{r^2_{R_l}} \end{bmatrix} \tag{2.22}$$

In the above formula, I_n is the n order unit matrix, q_n stands for the n matrix, in which its elements are all 1.

The geometric dilution of positioning (GDOP) accuracy is:

$$\mathrm{GDOP} = \sqrt{tr\left[P_{d\hat{X}}\right]}$$
$$= \sqrt{P_{d\hat{X}}(1,1) + P_{d\hat{X}}(2,2) + P_{d\hat{X}}(3,3)} \tag{2.23}$$

2.2.1.3 Estimated Cramér–Rao Bound of the Target Position of the Receiver Station

Set z_l to the measurements of $l (l = 1$ or $2, \ldots$ or $m)$ receiving station, where the receiving station in the range and azimuth and pitch angle measurement error are independent of each other, and conform to a Gaussian distribution with zero mean and their standard deviations are, respectively, $\delta_{\rho_{R_l T_1}}$, $\delta_{\varphi_{R_l}}$, and $\delta_{\varepsilon_{R_l}}$, the measurement noise between the receiving station is independent, and (2.1) can be rewritten as follows:

$$z_l = h_l(X) + w_l \tag{2.24}$$

$X = [x, y, z]^T$ is the target position vector, $w_l = \left[w_{\rho_{R_l T_1}}, \cdots, w_{\rho_{R_l T_n}}, w_{\varphi_{R_l}}, w_{\varepsilon_{R_l}}\right]^T$ is the measured noise of the l receiving station, and the covariance matrix is:

$$R_c = \begin{bmatrix} \delta_{\rho_{R_l T_1}}^2 & \eta_{12}\delta_{\rho_{R_l T_1}}\delta_{\rho_{R_l T_2}} & \cdots & \eta_{1n}\delta_{\rho_{R_l T_1}}\delta_{\rho_{R_l T_n}} & 0 & 0 \\ & \delta_{\rho_{R_l T_2}}^2 & \cdots & \eta_{2n}\delta_{\rho_{R_l T_2}}\delta_{\rho_{R_l T_n}} & 0 & 0 \\ - & & \ddots & & & \\ & & & \delta_{\rho_{R_l T_n}}^2 & 0 & 0 \\ 0 & 0 & \cdots & 0 & \delta_{\varphi_{R_l}}^2 & 0 \\ 0 & 0 & \cdots & 0 & 0 & \delta_{\varepsilon_{R_l}}^2 \end{bmatrix} \tag{2.25}$$

The likelihood function of the target position X is:

$$P[z_l|X] = \frac{1}{\sqrt{|2\pi R_c|}}\exp\left\{-\frac{1}{2}[z_l - h_l(X)]^T R_c^{-1}[z_l - h_l(X)]\right\} \tag{2.26}$$

From this, the Fisher information matrix of the target position estimation is obtained as:

$$J_l = E\left\{[\nabla_X \ln P(z_l|X)][\nabla_X \ln P(z_l|X)]^T\right\} \tag{2.27}$$

Among this equation:

$$\begin{aligned} \nabla_X \ln P[z_l|X] &= -\frac{d^T(z_l - h_l(X))}{dX}R_c^{-1}(z_l - h_l(X)) \\ &= -\left[\nabla_X \rho_{R_l T_1}, \cdots \nabla_X \rho_{R_l T_n}, \nabla_X \varphi_{R_l}, \nabla_X \varepsilon_{R_l}\right]R_c^{-1}(z_l - h_l(X)) \end{aligned}$$

Therefore:

$$E\left\{\left[\nabla_X \ln P(z_l|X)\right]\left[\nabla_X \ln P(z_l|X)\right]^T\right\}$$
$$= \left[\nabla_X \rho_{R_l T_1}, \cdots \nabla_X \rho_{R_l T_n}, \nabla_X \varphi_{R_l}, \nabla_X \varepsilon_{R_l}\right] R_c^{-1} E\left\{w_l w_l^T\right\} R_c^{-1} \left[\nabla_X \rho_{R_l T_1}, \cdots \nabla_X \rho_{R_l T_n}, \nabla_X \varphi_{R_l}, \nabla_X \varepsilon_{R_l}\right]^T$$
$$= \left[\nabla_X \rho_{R_l T_1}, \cdots, \nabla_X \rho_{R_l T_n}, \nabla_X \varphi_{R_l}, \nabla_X \varepsilon_{R_l}\right] R_c^{-1} \left[\nabla_X \rho_{R_l T_1}, \cdots, \nabla_X \rho_{R_l T_n}, \nabla_X \varphi_{R_l}, \nabla_X \varepsilon_{R_l}\right]^T$$

The final result is as shown in formula (2.26).

$$J_l = [\nabla_X] * [R_c]^{-1} * [\nabla_X]^T \tag{2.28}$$

Among this:

$$\nabla_X = -\left[\nabla_X \rho_{R_l T_1}, \cdots \nabla_X \rho_{R_l T_n}, \nabla_X \varphi_{R_l}, \nabla_X \varepsilon_{R_l}\right]$$

$$\nabla_X \rho_{R_l T_i} = \left[\frac{x-x_{R_l}}{r_{R_l}} + \frac{x-x_{T_i}}{r_{T_i}}, \frac{y-y_{R_l}}{r_{R_l}} + \frac{y-y_{T_i}}{r_{T_i}}, \frac{z-z_{R_l}}{r_{R_l}} + \frac{z-z_{T_i}}{r_{T_i}}\right]^T (i = 1, 2, \cdots, n)$$

$$\nabla_X \varphi_{R_l} = \left[-\frac{\sin^2 \phi_{R_l}}{y-y_{R_l}}, \frac{\cos^2 \phi_{R_l}}{x-x_{R_l}}, 0\right]^T$$

$$\nabla_X \varepsilon_{R_l} = \left[-\frac{c_{R_l 3} \cos \phi_{R_l}}{r_{R_l}}, -\frac{c_{R_l 3} \sin \phi_{R_l}}{r_{R_l}}, \frac{\cos \varepsilon_{R_l}}{r_{R_l}}\right]^T$$

So, the receiving station target location estimation Cramér–Rao bound P_{CRLB} is:

$$P_{\text{CRLB}} = J_l^{-1} \tag{2.29}$$

2.2.1.4 Simulation and Analysis

Simulation environment (two transmitters one receiver): The coordinates of transmitting station 1 are $T_1(-10, 0, 0.05)$km; the coordinates of transmitting station 2 are $T_2(10, 0, 0.05)$km; the coordinates of the receiving station are $R(30, 0, 0)$km. The target height is 10 and the scope of the target position is: x direction ± 200 km; y direction ± 200 km; the range and measurement error standard difference is 3 m; the location error standard deviation is 1 m; the azimuth and pitching angle measurement error standard deviation is 3 mrad ($0.17°$), and distance and correlation coefficient is $\eta = 0.3$. Compared with the positioning accuracy of the same one transmitter one emitter mode, the simulation results are shown in the following diagram:

Figure 2.1 is the GDOP distribution map located by transmitting station 1 only when it is operating. Comparison between Figs. 2.2 and 2.1 shows that the location accuracy of all the regions clearly increased, including the baseline region after increasing transmitting station 2.

As shown in Figs. 2.3 and 2.4, take $y = 10$ km and $y = 150$ km as examples. In the condition of two transmitters one receiver, the GDOP value approaches the Cramér–Rao bound, and this shows that the two transmitter one receiver positioning accuracy of the near field and far field in the y direction is very high.

Fig. 2.1 One transmitter
one receiver

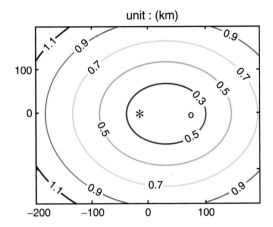

Fig. 2.2 Two transmitters
one receiver

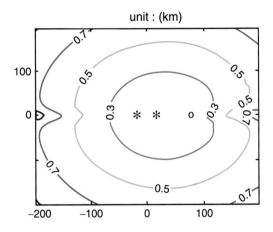

Fig. 2.3 Two transmitters
one receiver ($y = 10$ km)

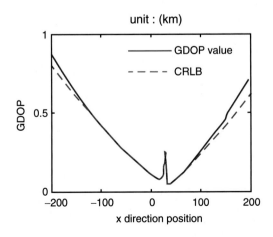

Fig. 2.4 Two transmitters
one receiver ($y = 150$ km)

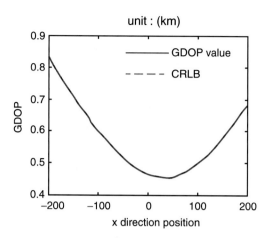

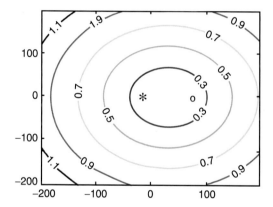

Fig. 2.5 One transmitter one receiver

If we increase the range and measuring error standard deviation to 10 m, and the other simulation conditions remain the same as simulation environment 1, the simulation results are shown as below in Fig. 2.8:

The comparison made between Figs. 2.6 and 2.2 shows that, if the range and standard error of the measurement error is increased from 3 to 10 m, the GDOP value of the baseline region is rapidly decreased, and the effect of the other regions is not obvious.

The comparison made between Figs. 2.6 and 2.5 shows that the baseline area of the extension of the online positioning accuracy is lower than the positioning accuracy of only one transmitter station when it is operating.

As shown in Figs. 2.7 and 2.8, taking the y direction as an example, the increase of the distance and the measurement error of the standard deviation leads to the fact that the GDOP value at $y = 10$ km is far from that of the Cramér–Rao bound, except

Fig. 2.6 Two transmitters
one receiver

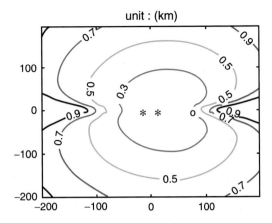

Fig. 2.7 Two transmitters
one receiver ($y = 10$ km)

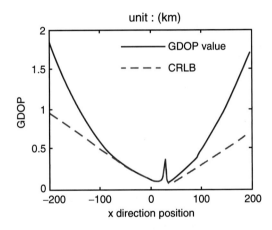

Fig. 2.8 Two transmitters
one receiver ($y = 150$ km)

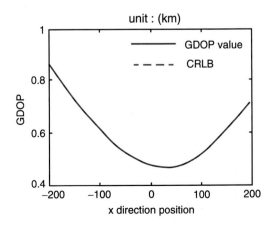

for a small area at the baseline, and the GDOP value at $y = 150$ km is still close to that of the Cramér–Rao bound. This shows that the increase in the range and measurement error standard difference will lead to the lower positioning accuracy of the near region in the y direction.

If the error standard of the azimuth measurement error is increased to 10 mrad (0.57°), and the other simulation conditions are the same as that in simulation environment 1, then the simulation results are as shown in the following diagrams:

Comparing Figs. 2.10 and 2.2, we can see that, if the azimuth measurement error standard error is increased from 3 to 10 mrad, it causes the baseline lateral extension line of the GDOP value to decrease rapidly. Comparing Figs. 2.9 and 2.10, we know that the positioning accuracy in these areas is even lower than the positioning accuracy when only one transmitting station is operating, and the influence on the positioning accuracy of the other regions is not obvious.

As shown in Figs. 2.11 and 2.12, taking the y direction as an example, the increase in the azimuth measurement error standard deviation will lead to the GDOP value at $y = 10$ km away from the Cramér–Rao bound, and the GDOP

Fig. 2.9 One transmitter one receiver

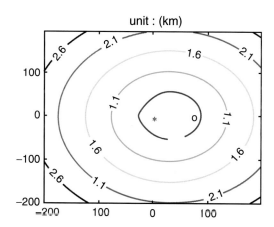

Fig. 2.10 Two transmitters one receiver

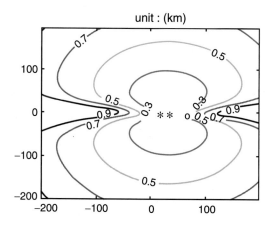

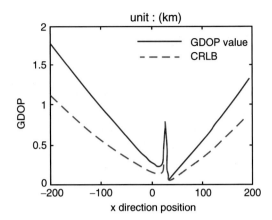

Fig. 2.11 Two transmitters one receiver ($y = 10$ km)

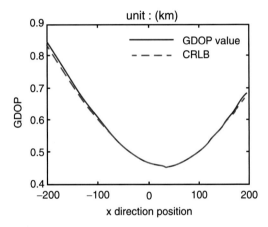

Fig. 2.12 Two transmitters one receiver ($y = 150$ km)

value at $y = 150$ km is close to the Cramér–Rao bound. This shows that the azimuth measurement error standard deviation causes an increase in the y direction positioning precision in near zone variation.

If the error standard of the pitch angle measurement is increased to 10 mrad (0.57°), and the other simulation conditions are the same as in that of simulation environment 1, then the simulation results are as shown in the following diagrams:

The comparison between Figs. 2.14 and 2.2 shows that, when the two transmitting stations are operating, if the pitch angle measurement standard error difference is increased from 3 to 10 mrad, this will result in a dramatic decrease in the positioning accuracy of each region. The GDOP value of many regions is increased to more than three times that of the original. The comparison between Figs. 2.14 and 2.13 shows that, in addition to the baseline lateral extension line, the regional positioning accuracy is higher than that when only one transmitting station is working, and the accuracy of the other regions is almost equal to the positioning accuracy of only one transmitting station when it is working.

As shown in Figs. 2.15 and 2.16, taking the y direction as an example, the increase in the pitch angle measurement error standard deviation leads to the GDOP

Fig. 2.13 One transmitter
one receiver

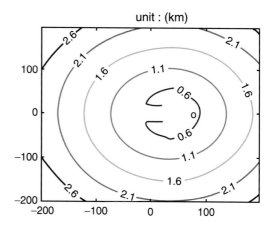

Fig. 2.14 Two transmitters
one receiver

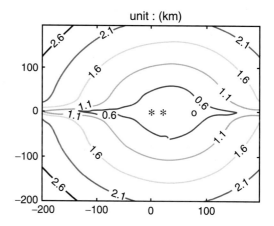

Fig. 2.15 Two transmitters
one receiver ($y = 10$ km)

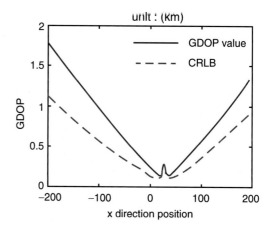

Fig. 2.16 Two transmitters
one receiver ($y = 150$ km)

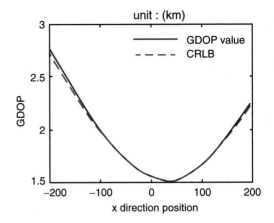

Fig. 2.17 One transmitter
one receiver

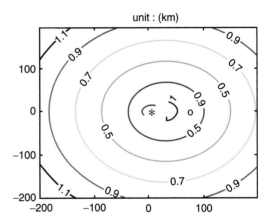

value at $y = 10$ km away from the Cramér–Rao bound, and the GDOP value at $y = 150$ km is close to the Cramér–Rao bound. This shows that the increase in the pitch angle measurement error standard deviation causes variation of the positioning accuracy in the y direction in the upper near zone.

If we increase the site error to 5 m and the other simulation conditions are the same as that in simulation environment 1, then the simulation results are shown as follow:

The comparison between Figs. 2.18 and 2.2 shows that the increase in the site standard error difference will greatly affect the positioning accuracy of the baseline lateral extension line region when two transmitting station are working, causing the GDOP value of the region to decrease rapidly outwards. The comparison between Figs. 2.18 and 2.17 shows that the positioning accuracy of the region is lower than the positioning accuracy of the corresponding region when only one transmitting station is operating.

Fig. 2.18 Two transmitters
one receiver

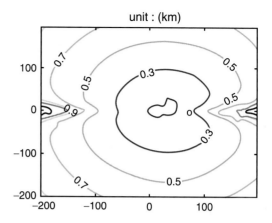

Fig. 2.19 Two transmitters
one receiver ($y = 10$ km)

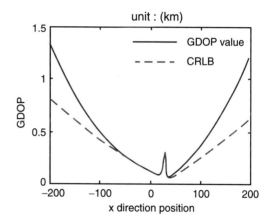

As shown in Figs. 2.19 and 2.20, in the y direction, as an example, the increase in the site standard error difference leads to the GDOP value at $y = 10$ km, far away from that of the Cramér–Rao bound, and the GDOP value at $y = 150$ km is still close to that of the Cramér–Rao bound. This shows that an increase in the site standard error difference will lead to variation of the positioning precision in the upper near zone in the y direction.

Conclusion When two transmitting stations are operating, the positioning accuracy of most of the regions of the target localization is significantly higher than that of the positioning accuracy when one transmitting station is operating. Changes in the range and measurement error and location error have little effect on the positioning accuracy. The influence of an increase in the azimuth measurement error on the positioning accuracy is greater than the effect of the distance and measurement error and location error on the positioning accuracy. The increase in the same pitch angle measurement error of the receiving station positioning

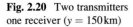
Fig. 2.20 Two transmitters
one receiver ($y = 150$ km)

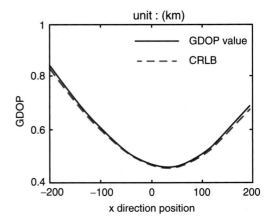

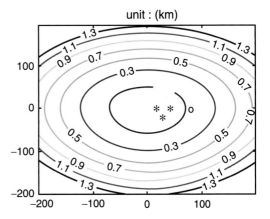
Fig. 2.21 Three
transmitters one receiver

accuracy is significantly greater than the increase in the same azimuth measurement error of the receiving station positioning accuracy.

Simulation environment 2 (three transmitters one receiver): Based on simulation environment 1, if we increase the number of transmitting stations $T_3(0, -10, 0.05)$ km, the simulation results are as shown in the following diagrams:

As shown in Fig. 2.21. when there are three transmitting stations, the GDOP curve will be approximately similar to an ellipse. In contrast, we can see from Fig. 2.2 that the positioning accuracy of the baseline of three transmitting stations is significantly improved The GDOP value, which is less than an area of 0.1 km, is increased significantly. Comparison among the results also show that the positioning accuracy in the elliptical semi-major axis direction corresponds to the positioning accuracy and short semi-axis direction of the regional positioning accuracy is lower than that of the two transmitting stations region, which is even lower than the

Fig. 2.22 Three transmitters one receiver ($y = 10$ km)

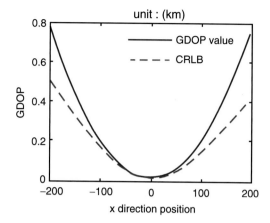

Fig. 2.23 Three transmitters one receiver ($y = 150$ km)

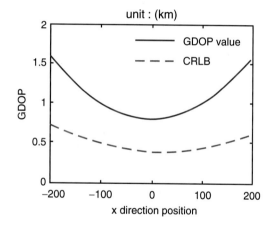

positioning accuracy of the corresponding region when one transmitting station is operating.

As shown in Figs. 2.22 and 2.23, in the y direction, for example, at $y = 10$ km, the positioning GDOP value is close to the Cramér–Rao bound, and at $y = 150$ km, the positioning GDOP value is far away from the Cramér–Rao bound. This shows that increasing the number of transmitting stations by one leads to the positioning accuracy in the y direction upper far zone to become worse.

Figure 2.24 shows similar results when the distance is increased and the measurement error standard deviation is increased to 5 m, while the other simulation conditions are the same as those in simulation environment 2. Figure 2.25 shows the simulation results when the site standard error difference is increased to 5 M and the other simulation conditions remain the same as those in simulation environment 2:

A comparison between Figs. 2.24 and 2.21 shows that, when the distance and measurement error of the standard deviation is increased from 3 to 5 m, the GDOP distribution curve remains as an ellipse, but the GDOP value of the long and short

Fig. 2.24 Three transmitters one receiver (range and measurement error standard deviation 5 m)

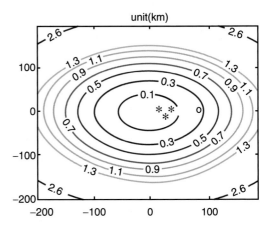

Fig. 2.25 Three transmitters one receiver (location error deviation 5 m)

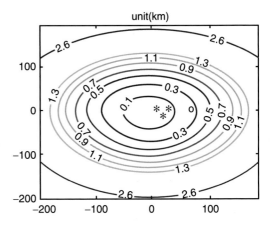

semi-axis direction far from the baseline region declined rapidly, especially in the semi minor axis direction, and the positioning performance is very poor..

A comparison between Figs. 2.25 and 2.21 shows that, when the distance and measurement error of the standard deviation is increased from 1 to 5 m, the GDOP distribution curve remains as an ellipse, but the GDOP value of the long and short semi-axis direction far from the baseline region declined rapidly, especially in the semi minor axis direction, and positioning performance is very poor.

Figure 2.26 shows the simulation results when the azimuth measurement error is increased to 10 mrad (0.57°), and the other simulation conditions are the same as those in simulation environment 2. Figure 2.27 shows the simulation results when the pitch angle error standard deviation is increased to 10 mrad (0.57°), and the other simulation conditions are the same as those in simulation environment 2.

Comparing Figs. 2.26, 2.27, and 2.21, we can see that, when the azimuth or pitch angle measurement error standard deviation is increased from 3 to 10 mrad, the positioning results are almost unaffected.

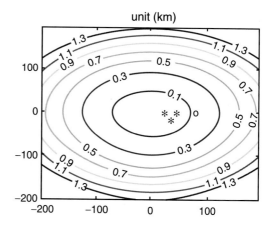

Fig. 2.26 Three transmitters one receiver (azimuth measurement error standard deviation 10 mrad)

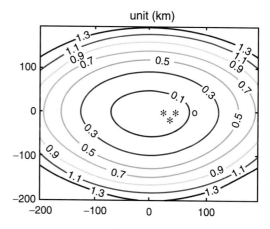

Fig. 2.27 Three transmitters one receiver (pitch angle measurement error standard deviation 10 mrad)

Simulation environment 3: Based on simulation environment 2, the number of transmitting stations is increased by one $T_4(0, 10, 0.05)$km. In the same simulation environment as in the similar simulation process of simulation environment 2, we get the same conclusion as that in simulation environment 2.

Conclusion When the transmitting station number is larger than or equal to three, the effect on the distance and measurement error and location error on the positioning accuracy is very obvious. However, the influence on changes in the azimuth angle and pitching angle measurement error on positioning precision is not obvious. This point is exactly the opposite to the conclusion arrived at in the condition where the number of transmitting stations is less than three. In addition, the positioning accuracy is not proportional to the number of launch stations, and an increase in the number of transmitting stations will only influence the positioning accuracy of the individual region.

2.2.2 One Transmitter Multiple Receiver Mode and Multiple Transmitter Multiple Receiver Mode

2.2.2.1 Distribution Centralized Structure

Fusion Positioning Algorithm of the Central Station

(a) The weighted least squares (WLS) estimation algorithm

Set the target location and its error of j receiving station for X_j and $dX_j (j = 1, 2, \cdots, m)$, respectively, then the m receiving station location subset corresponds to the position vector and the relationship between errors can be expressed as follows:

$$\overline{X} = HX + V \tag{2.30}$$

In this formula:

$$\overline{X} = \left[X_1^T, \cdots, X_m^T\right]^T$$

$$H = \left[I_1, \cdots, I_m\right]^T$$

$$V = \left[dX_1^T, \cdots, dX_m^T\right]^T$$

where:

1. $I_i (i = 1, 2, \cdots, m)$ and vector X is a dimensional unit matrix
2. $V \sim N(0, B)$, B is a km $\times$ km order matrix, k is the dimension of the target position vector X.

$$B = E\left[VV^T\right] = \begin{bmatrix} dX_1 X_1^T & dX_1 dX_2^T & \cdots & dX_1 dX_m^T \\ & dX_2 dX_2^T & \cdots & dX_2 dX_m^T \\ \underline{\qquad} & & \cdots & \cdots \\ & & & dX_m dX_m^T \end{bmatrix} \overset{\Delta}{=} \left[B_{ij}\right]_{m \times m} \tag{2.31}$$

$$B_{ij} \overset{\Delta}{=} E\left[dX_i dX_j^T\right] (i, j = 1, 2, \cdots, m) \tag{2.32}$$

B is a $k \times k$ order matrix, so the position subset of the m receiving station is a linear combination with the WLS algorithm, the target location estimation, and its error covariance matrix, respectively:

$$\hat{X}_{\text{WLS}} = \left(H^T B^{-1} H\right)^{-1} H^T B^{-1} \overline{X} \tag{2.33}$$

$$\hat{P}_{\text{WLS}} = \left(H^T B^{-1} H\right)^{-1} \tag{2.34}$$

Assuming:

$$B^{-1} \stackrel{\Delta}{=} G = \left[G_{ij} \right]_{m \times m} \tag{2.35}$$

$G_{ij}(i,j = 1, 2, \cdots, m)$ is a $k \times k$ order matrix, then:

$$\hat{X}_{\text{WLS}} = \left(\sum_{i=1}^{m} \sum_{j=1}^{m} G_{ij} \right)^{-1} \sum_{i=1}^{m} \sum_{j=1}^{m} G_{ij} X_j \tag{2.36}$$

$$\hat{P}_{\text{WLS}} = \left(\sum_{i=1}^{m} \sum_{j=1}^{m} G_{ij} \right)^{-1} \tag{2.37}$$

If the position subset of the m receiving station is not correlated, then:

$$B = \text{diag}[B_{11}, \cdots, B_{mm}] \tag{2.38}$$
$$G = \text{diag}\left[B_{11}^{-1}, \cdots, B_{mm}^{-1} \right] \tag{2.39}$$

Then:

$$\hat{X}_{\text{WLS}} = \left(\sum_{i=1}^{m} B_{ii}^{-1} \right)^{-1} \sum_{i=1}^{m} B_{ii}^{-1} X_i \tag{2.40}$$

$$\hat{P}_{\text{WLS}} = \left(\sum_{i=1}^{m} B_{ii}^{-1} \right)^{-1} \tag{2.41}$$

(b) Selection of the optimal WLS estimation fusion positioning algorithm

In general, when a small number of receiving stations are in fusion (less than or equal to 5), the correlation between each receiving location set can be ignored. The application type in formula (2.36) can give good fusion results and higher operation efficiency. But when there are greater numbers of receiving stations in fusion (more than five), on the one hand the possibility of correlation between the position sets of some stations is increased significantly, and some of them cannot even be ignored, the WLS algorithm involves the mutual covariance matrix of the position set of every group, with km × km high-order inverse matrix of complex calculations, so it is very difficult to apply. On the other hand, if there are more data transmitted to the central station, not only is great communication bandwidth required, the central station needs to have strong data processing capabilities, which increases the complexity of the system, and also the timeliness of the fusion location of the central station is reduced, especially as the location sets of some receiving stations involved in fusion have a smaller contribution to the fusion results of the central station. In response to these problems, when the number of receiving stations in fusion is large (more than five), we put forward the selection of the optimal WLS estimation fusion location algorithm of the central station.

The detailed steps of the algorithm are as follows:

1. Using the subset selection method to select the five location sets of the receiving station with the best positioning accuracy.

   Set the location subset of j receiving station corresponding to the target location and error covariance matrix as, respectively, X_j and $B_j(j = 1, 2, \cdots, m)$. At every point in a controlled area, the comparison of the variable $\gamma_j = [\det B_j]^{1/2}$ is carried out in order to select the target location data corresponding to the position set of five minimum $\gamma^* = \min\{\gamma_j, j = 1, 2, \cdots, m\}$. That is:

   If:

$$[\det B_j]^{1/2} = \min\left\{[\det B_1]^{1/2}, \cdots, [\det B_m]^{1/2}\right\} \qquad (2.42)$$

   Then:

$$\begin{cases} \hat{X}_{opt} = X_j, & \text{Target position estimation} \\ \hat{P}_{opt} = B_j, & \text{Error covariance matrix} \end{cases} \qquad (2.43)$$

2. The position set of the five receiving stations obtained by formula (2.39) and the error covariance matrix s are sent to the central station. Use formula (2.36) for fusion processing to reach the final fusion value, and the corresponding error covariance matrix is the same as that in formula (2.37).

The Central Station Estimates the Target Location in the Cramér–Rao Bound

Suppose that Z represents a collection of measurements of m receiving stations sent to the network central station. That is, $Z = \{z_l\}_{l=1}^m$, z_l as shown in formula (2.23). Assuming that the measurement noise is independent, therefore, the likelihood function of target position X of the central station is c:

$$P[Z|X] = P[z_1, z_2, \cdots, z_m|X]$$
$$= \prod_{l=1}^{m} P[z_l|X] \qquad (2.44)$$

The Fisher information matrix of the central station target position of the estimation is:

$$J = E\left\{[\nabla_X \ln P(Z|X)][\nabla_x \ln P(Z|X)]^T\right\} \qquad (2.45)$$

By formulas (2.40) and (2.41), we get:

$$J = J_1 + J_2 + \cdots + J_m \tag{2.46}$$

In this equation, $J_l(l = 1, 2 \cdots, m)$, as shown in formula (2.26).

So, the Cramér–Rao bound of the network center target position estimation P_{CRLB} is shown by the following formula:

$$P_{CRLB} = (J_1 + J_2 + \cdots + J_m)^{-1} \tag{2.47}$$

Simulation and Analysis

(a) One transmitter multiple receiver mode

In the previous section, we arrived at the general conclusion about the influence of the changes in various measurement errors on the positioning accuracy of each individual receiving station. In this section, we will focus on the influence of changes in various errors on the positioning accuracy of the central station.

Simulation environment 4 (one transmitter four receivers): the transmitting station coordinates are $T(-10, 0, 0.05)$ km; receiving station 1 coordinates are $R_1(30, 0, 0)$ km; receiving station 2 coordinates are $R_2(0, 30, 0)$ km; receiving station 3 coordinates are $R_3(0, -30, 0)$ km ; receiving station 4 coordinates are $R_4(-30, 0, 0)$ km. The target height is 10 km and the range of target location is: x direction ± 200 km, y direction ± 200 km, the range and measurement error standard deviation is 3 m, site error is 1 m, range and correlation coefficient $\eta = 0.3$. The azimuth and elevation measurement error standard deviation is 3 mrad (0.17°). The simulation results are shown in Figs. 2.28, 2.29, 2.30, 2.31, 2.32, 2.33, and 2.34.

Comparing Fig. 2.32 and Figs. 2.31, 2.32, 2.33, 2.34, 2.35, 2.36, 2.37, and 2.38, we can see that, after the network center station conducts the fusion process for the location of each receiving station, the positioning accuracy was significantly higher

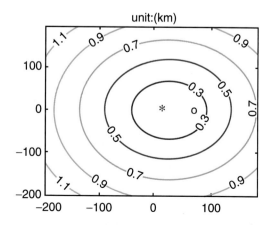

Fig. 2.28 Receiving station 1

Fig. 2.29 Receiving
station 2

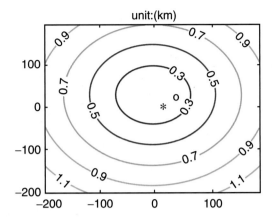

Fig. 2.30 Receiving
station 3

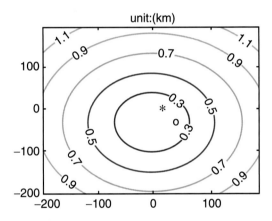

Fig. 2.31 Receiving
station 4

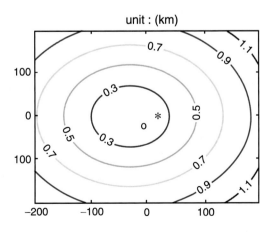

Fig. 2.32 Distribution:
centralized central station

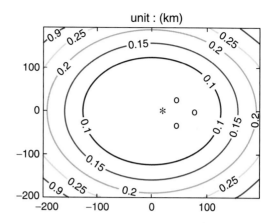

Fig. 2.33 Distribution:
centralized central station
$y = 10\,\mathrm{km}$

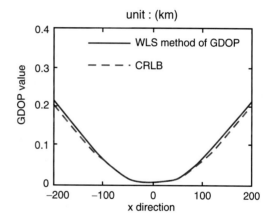

Fig. 2.34 Distribution:
centralized central station
$y = 150\,\mathrm{km}$

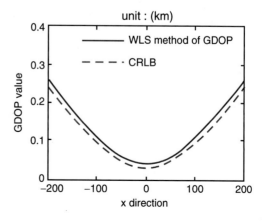

Fig. 2.35 Distribution: centralized central station

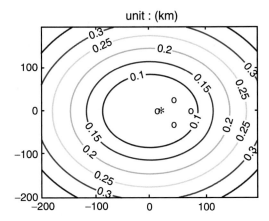

Fig. 2.36 GDOP of Fig. 2.35 Distribution: centralized central station

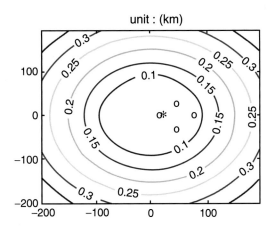

Fig. 2.37 Distribution: centralized central station (azimuth measurement error standard deviation 10 mrad)

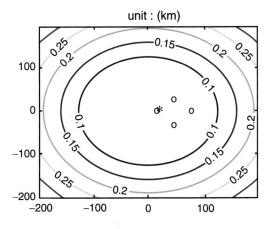

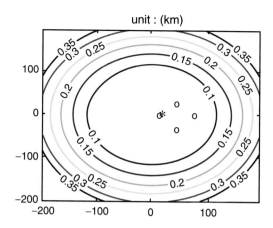

Fig. 2.38 Distribution: centralized central station (pitch angle measurement error standard deviation 10 mrad)

than that of any receiving station, especially when considering that the improvement of the positioning accuracy at the baseline is very obvious, and the other GDOP values are about a quarter of the corresponding region of each receiving station.

As shown in Figs. 2.33 and 2.34, taking the y direction as an example, no matter whether in the far range $y = 150$ km or in the near range $y = 10$ km, the GDOP values of the central fusion positioning are approximately close to that of the Cramér–Rao bound.

Figure 2.35 shows the GDOP distribution of the central station when the range and measurement error standard deviation is increased to 10 m, and the other simulation conditions are the same as those in simulation environment 4. Figure 2.36 shows the GDOP distribution of the central station when the range and measurement error standard deviation is increased to 5 m, and the other simulation conditions are the same as those in simulation environment 4.

A comparison between Figs. 2.35 and 2.32 shows that the increase in the range measurement error and standard deviation leads to a significant decrease of the central station positioning accuracy, and the GDOP value of an area less than 0.1 km was significantly reduced.

Figure 2.37 shows the effect of increasing the azimuth measurement error to the standard deviation 10 mrad (0.57°), while the other conditions of the simulation remained the same as those of the GDOP value of the central station of simulation environment 4.

Figure 2.38 shows the effect of increasing the pitch angle measurement error to the standard deviation 10 mrad (0.57°), while the other conditions of the simulation remained the same as those of GDOP value of the central station of simulation environment 4.

A comparison between Figs. 2.37 and 2.32 shows that, when the azimuth measurement error standard deviation is increased from 3 to 10 mrad, there was no effect on the positioning accuracy of the center station.

Fig. 2.39 Distribution: centralized central station (fusion of four stations)

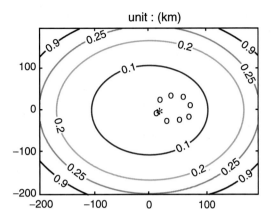

Fig. 2.40 Distribution: centralized central station (fusion of five stations)

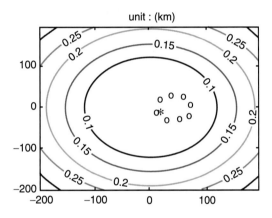

A comparison between Figs. 2.38 and 2.32 shows that, when the elevation angle measurement error standard deviation is increased from 3 to 10 mrad, it lead to significant changes in the positioning accuracy of the central station. The contour line of the GDOP value greater than 0.1 km dropped away faster. On the basis of the one transmitter four receiver mode, increasing the number of receiving stations with receiving station 5 having coordinates $R_5(21.2, 21.2, 0)$km, receiving station 6 having coordinates $R_6(-21.2, 21.2, 0)$km, receiving station 7 having coordinates $R_7(-21.2, -21.2, 0)$km, and receiving station 8 having coordinates $R_8(21.2, -21.2, 0)$km, the simulation results are as shown in the following figure:

As shown in Figs. 2.42 and 2.39, the positioning result of the selected WLS estimation fusion location algorithm for the set of five station locations is closer to the optimal value obtained by the set of eight station locations as a whole; and Fig. 2.41 shows that, despite the fusion of the position sets of six receiving stations, in some areas, the positioning accuracy was significantly lower than that in the corresponding region in Fig. 2.40. This shows the selection of the optimal WLS estimation fusion localization algorithm is effective and reasonable.

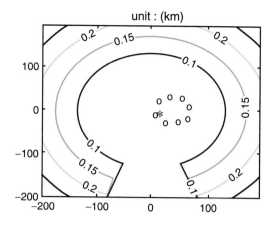

Fig. 2.41 Distribution: centralized central station (fusion of six stations)

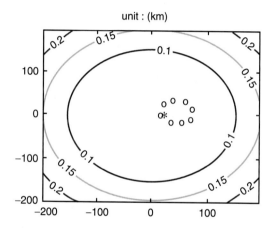

Fig. 2.42 Distribution: centralized central station (fusion of eight stations)

Conclusion In one transmitter multiple receiver mode, the increase in the pitch angle measurement error causes the central station positioning accuracy to decrease significantly; while the increase in the range and measurement error and location error will lead to a significant decrease of the central station positioning accuracy. The central station positioning accuracy is less affected by the azimuth measurement error. When the number of the receiving stations is more than five, the selection of the optimal WLS estimation fusion location algorithm produced good results.

(b) Multiple transmitter multiple receiver mode

Simulation environment 5 (four transmitters four receivers): the coordinates of transmitting station 1 are $T_1(-10, 0, 0.05)$km, the coordinates of transmitting station 2 are $T_2(10, 0, 0.05)$km, the coordinates of transmitting station 3 are $T_3(0, -10, 0.05)$km, the coordinates of transmitting station 4 are $T_4(0, 10, 0.05)$km, the coordinates of receiving station 1 are $R_1(30, 0, 0)$km, the coordinates of

Fig. 2.43 Receiving station 1

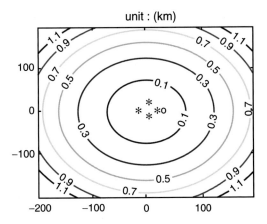

Fig. 2.44 Receiving station 2

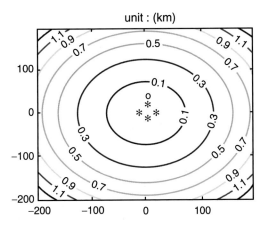

receiving station 2 are $R_2(0, 30, 0)$km, the coordinates of receiving station 3 are $R_3(0, -30, 0)$km, the coordinates of receiving station 4 are $R_4(-30, 0, 0)$km. The target height is 10 km, the range of target location is : x direction ± 200 km, y direction ± 200 km, range and measurement error standard deviation 3 m, location error 1 m, the range and correlation coefficient $\eta = 0.3$, and azimuth and elevation measurement error standard deviation 3 mrad (0.17°). The simulation results are shown in Figs. 2.43, 2.44, 2.45, 2.46, 2.47, 2.48, and 2.49.

Comparison of Fig. 2.47 and Figs. 2.43, 2.44, 2.45, and 2.46 shows that, after the network center station conducts the fusion process for the location data of each receiving station, the target position estimation accuracy was higher than that of any receiving station, and the corresponding GDOP value is about 1/4 for each receiving station.

As shown in Figs. 2.48 and 2.49, taking the y direction as an example, no matter whether in the far range $y = 150$ km or in the near range $y = 10$ km, the GDOP values of the central fusion positioning is close to that of the Cramér–Rao bound.

Fig. 2.45 Receiving
station 3

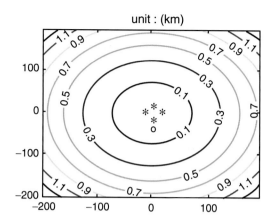

Fig. 2.46 Receiving
station 4

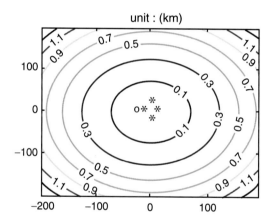

Fig. 2.47 Distribution:
centralized central station

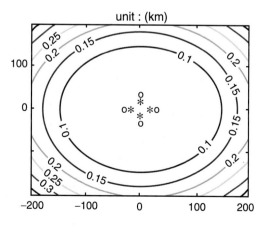

Fig. 2.48 Distribution: centralized central station ($y = 10$ km)

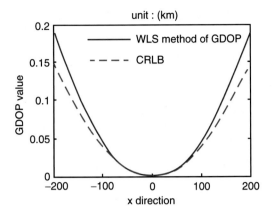

Fig. 2.49 Distribution: centralized central station ($y = 150$ km)

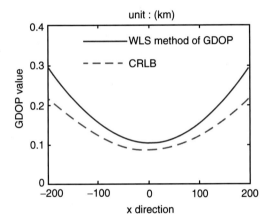

Figure 2.50 shows the GDOP figures of the central station when the range and measurement error standard deviation is increased to 10 m, and the other conditions of the simulation are the same as those in simulation environment 5. Figure 2.51 shows the GDOP figure of the central station when the range and measurement error standard deviation is increased to 5 m, and the other conditions of the simulation are the same as those in simulation environment 5.

A comparison between Figs. 2.51 and 2.47 shows that, when the range and measurement error standard deviation is increased from 1 to 5 m, it also leads to significant changes in the positioning accuracy of the central station, and the range of GDOP value which is less than 0.1 km is decreased obviously, and the GDOP curve is decreased rapidly outwards.

A comparison of Figs. 2.51 and 2.47 shows that, when the location error is increased from 1 to 5 m, it also leads to the positioning accuracy of the central station being significantly reduced.

Fig. 2.50 Distribution: centralized central station (range and measurement error of standard deviation 10 m)

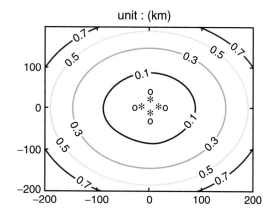

Fig. 2.51 Distribution: centralized central station (standard deviation of site error 5 m)

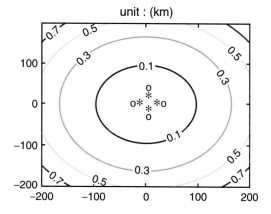

Figure 2.52 shows that the results when the azimuth measurement error to standard deviation is increased to 10 mrad (0.57°) and the other conditions of the simulation are the same as those of the GDOP figure of the central station of simulation environment 5. Figure 2.52 shows the results when the pitch angle measurement error to standard deviation is increased to 10 mrad (0.57°) and the other conditions of the simulation are the same as those of the GDOP figure of the central station of simulation environment 5:

A comparison of Figs. 2.52 and 2.47 shows that, when the azimuth measurement error standard deviation is increased from 3 to 10 mrad, the positioning accuracy of the central station was hardly changes.

A comparison of Figs. 2.53 and 2.47 shows that, when the pitch angle measurement error standard deviation is increased from 3 to 10 mrad, the positioning accuracy of the central station is not affected.

Simulation environment 6 (four transmitters eight receivers): On the basis of simulation environment 5, increasing the number of receiving stations with receiving station 5 having coordinates $R_5(21.2, 21.2, 0)$km, receiving station 6 having coordinates $R_6(-21.2, 21.2, 0)$ km, receiving station 7 having coordinates

Fig. 2.52 Distribution:
centralized central station
(azimuth measurement error
standard deviation 10 mrad)

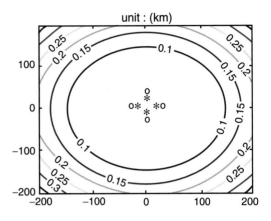

Fig. 2.53 Distribution –
centralized central station
(pitch angle measurement
error of standard deviation
10 mrad)

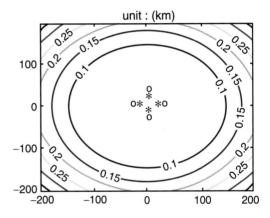

$R_7(-21.2, -21.2, 0)$ km, and receiving station 8 having coordinates $R_8(21.2, -21.2, 0)$ km, the simulation results are shown in the following figures.

As shown in Figs. 2.54, 2.55, and 2.56, the selection of the optimal WLS estimation fusion location algorithm shows little difference from the fusion results of fusing the location data of the five receiving stations and six receiving stations. The results are close to the fusion of the eight stations' location data, and it shows that, when the number of receiving stations is more than five, only five receiving stations' locations has the highest positioning accuracy of transmission to the central station set, and it can obtain good results.

Conclusion In multiple receiver multiple transmitter mode, the increase in the range and the measurement error and the position error will lead to a significant decrease of the positioning accuracy of the central station. The increase in the azimuth and elevation error has little influence on the positioning accuracy of the central station. When the number of receiving stations is more than five, only five receiving stations' locations has the highest positioning accuracy of transmission to the central station set, and it can obtain good results.

Fig. 2.54 Distribution:
centralized central station
(fusion of five stations)

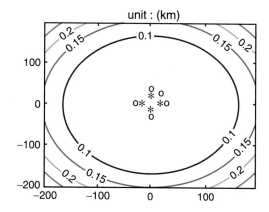

Fig. 2.55 Distribution:
centralized central station
(fusion of six stations)

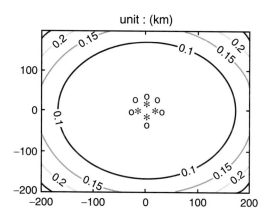

Fig. 2.56 Distribution:
centralized central station
(fusion of eight stations)

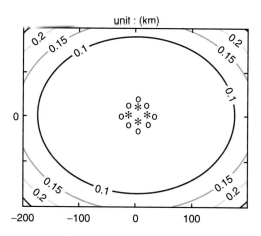

2.2.2.2 Centralized Structure

Acquisition of the Positioning Solution

According to the observed quantity $\rho_{R_l T_i}$ acquired by station $R_l, (l = 1, 2, \cdots, m)$, we can obtain the following equation:

$$
\begin{cases}
r_{T_i} = \sqrt{(x - x_{T_i})^2 + (y - y_{T_i})^2 + (z - z_{T_i})^2} \, (i = 1, 2, \cdots, n) \\
r_{R_l} = \sqrt{(x - x_{R_l})^2 + (y - y_{R_l})^2 + (z - z_{R_l})^2} \\
\rho_{R_l T_i} = r_{R_l} + r_{T_i} \, (l = 1, 2, \cdots, m)
\end{cases}
\tag{2.48}
$$

By transposition, square, finishing simplification of the formula, we get:

$$
(x_{T_i} - x_{R_l})x + (y_{T_i} - y_{R_l})y + (z_{T_i} - z_{R_l})z = k_{R_l T_i} - \rho_{R_l T_i} r_{T_i}
\tag{2.49}
$$

In this equation, $k_{R_l T_i} \triangleq \frac{1}{2} \left[\rho_{R_l T_i}^2 + \left(x_{T_i}^2 + y_{T_i}^2 + z_{T_i}^2 \right) - \left(x_{R_l}^2 + y_{R_l}^2 + z_{R_l}^2 \right) \right]$ $m \times n$. The equation expressed in formula (2.45) constitutes a non-linear equation. We first take the indirect method, the $r_{T_i} (i = 1, \cdots, n)$ as a known quantity, then we get that x, y, z is a function of $r_{T_i} (i = 1, \cdots, n)$, and apply x, y, z into $r_{T_i} (i = 1, \cdots, n)$. Then, we get the expression $r_{T_i} (i = 1, \cdots, n)$ and the specific process is as follows.

The type (2.45) specific forms of the $m \times n$ equation are written as follows:

$$
\begin{bmatrix}
x_{T_1} - x_{R_1} & y_{T_1} - y_{R_1} & z_{T_1} - z_{R_1} \\
\vdots & \vdots & \vdots \\
x_{T_1} - x_{R_m} & y_{T_1} - y_{R_m} & z_{T_1} - z_{R_m} \\
\vdots & \vdots & \vdots \\
x_{T_n} - x_{R_1} & y_{T_n} - y_{R_1} & z_{T_n} - z_{R_1} \\
\vdots & \vdots & \vdots \\
x_{T_n} - x_{R_m} & y_{T_n} - y_{R_m} & z_{T_n} - z_{R_m}
\end{bmatrix}
\begin{bmatrix} x \\ y \\ z \end{bmatrix}
=
\begin{bmatrix}
k_{R_1 T_1} - \rho_{R_1 T_1} r_{T_1} \\
\vdots \\
k_{R_m T_1} - \rho_{R_m T_1} r_{T_1} \\
\vdots \\
k_{R_1 T_n} - \rho_{R_1 T_n} r_{T_n} \\
\vdots \\
k_{R_m T_n} - \rho_{R_m T_n} r_{T_n}
\end{bmatrix}
\tag{2.50}
$$

Or can be written as:

$$
BX = g
\tag{2.51}
$$

Selecting the appropriate location rank (b) = 3 and solving equations by the use of the pseudo inverse method (2.47), we get:

$$
\hat{X} = (B^T B)^{-1} B^T g
\tag{2.52}
$$

Assuming that:

$$
(B^T B)^{-1} B^T \triangleq
\begin{bmatrix}
b_{11} & \cdots & b_{1(m \times n)} \\
b_{21} & \cdots & b_{2(m \times n)} \\
b_{31} & \cdots & b_{3(m \times n)}
\end{bmatrix}
= [b_{ij}]_{3 \times (m \times n)}
\tag{2.53}
$$

by formula (2.48) we get:

$$
\begin{cases}
\hat{x} = \sum_{i=1}^{n}\sum_{j=1}^{m} b_{1[(i-1)\times m+j]} k_{R_jT_i} - \sum_{i=1}^{n}\sum_{j=1}^{m}\left(b_{1[(i-1)\times m+j]}\rho_{R_jT_i}\right)r_{T_i} \overset{\Delta}{=} m_1 - n_1 r_{T_i} \\
\hat{y} = \sum_{i=1}^{n}\sum_{j=1}^{m} b_{2[(i-1)\times m+j]} k_{R_jT_i} - \sum_{i=1}^{n}\sum_{j=1}^{m}\left(b_{2[(i-1)\times m+j]}\rho_{R_jT_i}\right)r_{T_i} \overset{\Delta}{=} m_2 - n_2 r_{T_i} \\
\hat{z} = \sum_{i=1}^{n}\sum_{j=1}^{m} b_{3[(i-1)\times m+j]} k_{R_jT_i} - \sum_{i=1}^{n}\sum_{j=1}^{m}\left(b_{3[(i-1)\times m+j]}\rho_{R_jT_i}\right)r_{T_i} \overset{\Delta}{=} m_3 - n_3 r_{T_i}
\end{cases}
$$

$$(2.54)$$

Among them:

$$
\begin{cases}
m_l = \sum_{i=1}^{n}\sum_{j=1}^{m} b_{l[(i-1)\times m+j]} k_{R_jT_i} \\
n_l = \sum_{i=1}^{n}\sum_{j=1}^{m}\left(b_{l[(i-1)\times m+j]}\rho_{R_jT_i}\right)
\end{cases}
\quad (l = 1, 2, 3) \qquad (2.55)
$$

Apply type (2.50) into the expressions r_{T_i}, we get:

$$
a r_{T_i}^2 - 2b r_{T_i} + c = 0 \qquad (2.56)
$$

In the above formula:

$$
\begin{cases}
a = n_1^2 + n_2^2 + n_3^2 - 1 \\
b = n_1(m_1 - x_{T_i}) + n_2(m_2 - y_{T_i}) + n_3(m_3 - z_{T_i}) \\
c = (m_1 - x_{T_i})^2 + (m_2 - y_{T_i})^2 + (m_3 - z_{T_i})^2
\end{cases}
\qquad (2.57)
$$

By (2.52), we can solve the r_{T_i} values of r_{T_i1} and r_{T_i2}, so there exists a locating ambiguity. If for A and B one is positive and one is negative, then we take the positive value. Otherwise, it can be combined with range of azimuth parameters. The specific methods are as follows: the r_{T_i1} and r_{T_i2} are applied into formula (2.50) and we get the target location estimation of two points $(\hat{x}_1, \hat{y}_1, \hat{z}_1)$ and $(\hat{x}_2, \hat{y}_2, \hat{z}_2)$, and the two azimuths $\hat{\theta}_1$ and $\hat{\theta}_2$ of receiving station A R_l can be obtained by calculation. If $|\hat{\theta}_1 - \theta_{R_l}| \leq |\hat{\theta}_2 - \theta_{R_l}|$, then we take $\hat{X} = [\hat{x}_1, \hat{y}_1, \hat{z}_1]^T$, otherwise, $\hat{X} = [\hat{x}_2, \hat{y}_2, \hat{z}_2]^T$, thus eliminating the location ambiguity.

Locating Error Analysis

For the $\rho_{R_lT_i} = r_{T_i} + r_{R_l}$ differential on both sides:

$$dp_{R_lT_i} = (c_{T_i1} + c_{R_l1})dx + (c_{T_i2} + c_{R_l2})dy + (c_{T_i3} + c_{R_l3})dz \\ + k_{T_i} + k_{R_l}(i = , 1, 2, \cdots, n; \ l = 1, 2, \cdots, m) \tag{2.58}$$

In the formula $c_{ip}(j = T_i$ or $R_i, p = 1, 2, 3)$, with formula (2.12):

$$\begin{cases} k_{T_i} \triangleq -\left[c_{T_i1}dx_{T_i} + c_{T_i2}dy_{T_i} + c_{T_i3}dz_{T_i}\right] \\ k_{R_l} \triangleq -\left[c_{R_l1}dx_{R_l} + c_{R_l2}dy_{R_l} + c_{R_l3}dz_{R_l}\right] \end{cases} \tag{2.59}$$

The $m \times n$ error equation which is shown in formula (2.54) is written in the form of a vector matrix:

$$dY = CdX + dX_s \tag{2.60}$$

In the above formula, $\begin{aligned} dY &= \left[d\rho_{R_1T_1}, \cdots, d\rho_{R_mT_1}, \cdots, d\rho_{R_1T_n}, \cdots, d\rho_{R_mT_n}\right]^T \\ dX &= [dx, dy, dz]^T \\ dX_s &= \left[k_{T_1} + k_{R_1}, \cdots, k_{T_1} + k_{R_m}, \cdots, k_{T_n} + k_{R_1}, \cdots, k_{T_n} + k_{R_m}\right]^T \end{aligned}$

$$C = \begin{bmatrix} c_{T_11} + c_{R_11} & c_{T_12} + c_{R_12} & c_{T_13} + c_{R_13} \\ \vdots & \vdots & \vdots \\ c_{T_11} + c_{R_m1} & c_{T_12} + c_{R_m2} & c_{T_13} + c_{R_m3} \\ \vdots & \vdots & \vdots \\ c_{T_n1} + c_{R_11} & c_{T_n2} + c_{R_12} & c_{T_n3} + c_{R_13} \\ \vdots & \vdots & \vdots \\ c_{T_n1} + c_{R_m1} & c_{T_n2} + c_{R_m2} & c_{T_n3} + c_{R_m3} \end{bmatrix}$$

Using the pseudo-inverse method, we can solve the locating error of target estimation by formula (2.56):

$$d\hat{X} = (C^TC)^{-1}C^T[dY - dX_s] \tag{2.61}$$

assuming that:

$$(C^TC)^{-1}C^T \triangleq B \triangleq [b_{ij}]_{3 \times m} \tag{2.62}$$

Because the transmitting station and measuring range of each network station's radar countermeasure system contain the timing system delivery error, station R_l brings the timing error by itself, as well as the target scattering delay jitter error. So the range and data measured by the m station contains some common error factors. Therefore, when $i = j, l = p$, the observation error $\rho_{R_lT_i}$ and $\rho_{R_pT_j}$ are related. By setting the range and measurement error of the system to be revised to zero mean, the location error in each measurement is unchanged, and that between the site error of each component of the $\left(dx_j, dy_j, dz_j\right) (j = T_i$ or $R_i)$ and each site error are uncorrelated. Therefore, the covariance of the locating error is $\left(\because E\left[d\hat{X}\right] = 0\right)$:

$$
\begin{aligned}
P_{d\hat{X}} &= E\left[d\hat{X}\, d\hat{X}^T\right] \\
&= B\left\{E\left[dY dY^T\right] + E\left[dX_s dX_s^T\right]\right\}B^T
\end{aligned}
\tag{2.63}
$$

In the formula:

$$
E\left[dY dY^T\right] = \begin{bmatrix}
\delta_{\rho_{11}}^2 & \eta_T \delta_{\rho_{11}}\delta_{\rho_{21}} & \cdots & \eta_T \delta_{\rho_{11}}\delta_{\rho_{m1}} & \eta_R \delta_{\rho_{11}}\delta_{\rho_{12}} & \cdots & 0 & \cdots & \eta_R \delta_{\rho_{11}}\delta_{\rho_{1n}} & \cdots & 0 \\
\eta_R \delta_{\rho_{m1}}\delta_{\rho_{11}} & \ddots & & \delta_{\rho_{m1}}^2 & 0 & \cdots & \eta_R \delta_{\rho_{m1}}\delta_{\rho_{m2}} & \cdots & 0 & \cdots & \eta_R \delta_{\rho_{m1}}\delta_{\rho_{mn}} \\
\eta_R \delta_{\rho_{1n}}\delta_{\rho_{11}} & \cdots & & 0 & \cdots & \eta_R \delta_{\rho_{1n}}\delta_{\rho_{1n-1}} & \cdots & 0 & \delta_{\rho_{1n}}^2 & \eta_T \delta_{\rho_{1n}}\delta_{\rho_{2n}} & \cdots & \eta_T \delta_{\rho_{1n}}\delta_{\rho_{mn}} \\
0 & \cdots & \eta_R \delta_{\rho_{mn}}\delta_{\rho_{m1}} & \cdots & 0 & \cdots & \eta_R \delta_{\rho_{mn}}\delta_{\rho_{mn-1}} & \eta_T \delta_{\rho_{mn}}\delta_{\rho_{1n}} & \eta_T \delta_{\rho_{mn}}\delta_{\rho_{2n}} & \cdots & \delta_{\rho_{mn}}^2
\end{bmatrix}
$$

In the above, $\delta_{\rho_{ij}} (i = 1, 2, \cdots, n; j = 1, 2, \cdots, m)$ represents the standard deviation of the sum of the measured target range between the i receiving station and the j transmitting station, η_T represents the correlation coefficient of the mutual transmitting station to sum of the ranges measured by the receiving station. η_R represents the correlation coefficient of the sum of the ranges to different transmitting stations and mutual receiving stations. They are defined as follows:

$$
\begin{aligned}
\eta_T &= \frac{\mathrm{cov}\left(d\rho_{T_i R_l}, d\rho_{T_i R_j}\right)}{\delta_{\rho_{il}}\delta_{\rho_{ij}}} \\
\eta_R &= \frac{\mathrm{cov}\left(d\rho_{T_i R_i}, d\rho_{T_j R_i}\right)}{\delta_{\rho_{li}}\delta_{\rho_{ji}}}
\end{aligned}
\tag{2.64}
$$

We can assuming that the site error of each component of the standard deviation is the same, and $\delta_{xj}^2 = \delta_{yj}^2 = \delta_{zj}^2 = \delta_s^2$, because $c_{j1}^2 + c_{j2}^2 + c_{j3}^2 = 1 (j = T_i$ or $R_l)$, to get:

$$
E\left[dX_s dX_s^T\right] = \begin{bmatrix}
2\delta_s^2 & \delta_s^2 & \cdots & \delta_s^2 & \delta_s^2 & \cdots & 0 & \cdots & \delta_s^2 & \cdots & 0 \\
\delta_s^2 & \ddots & \cdots & \delta_s^2 & 2\delta_s^2 & 0 & \cdots & \delta_s^2 & \cdots & 0 & \cdots & \delta_s^2 \\
\delta_s^2 & \cdots & 0 & \cdots & \delta_s^2 & \cdots & 0 & 2\delta_s^2 & \delta_s^2 & \cdots & \delta_s^2 \\
0 & \cdots & \delta_s^2 & \cdots & 0 & \cdots & \delta_s^2 & \delta_s^2 & \cdots & \delta_s^2 & 2\delta_s^2
\end{bmatrix}
\tag{2.65}
$$

Location accuracy:

$$\begin{aligned} \text{GDOP} &= \sqrt{tr\left[P_{d\hat{X}}\right]} \\ &= \sqrt{P_{d\hat{X}}(1,1) + P_{d\hat{X}}(2,2) + P_{d\hat{X}}(3,3)} \end{aligned} \tag{2.66}$$

The Cramér–Rao Bound of the Target Location Estimation of the Central Station

Suppose Z is a collection of the measured values of the center station, with measurement error of each receiving station conforming to a Gaussian distribution with zero mean, then $\delta_{\rho_{ij}}(i=1,2,\cdots,n; j=1,2,\cdots,m)$ represents the standard deviation of the sum of the measured target ranges between i receiving station and j transmitting station. Because the transmitting station and measuring range of each network station's radar countermeasure system contain the timing system delivery error, the R_l station brings the timing error by itself, as well as the target scattering delay jitter error. So, range and data measured by the m station contained some common error factors; therefore, $\rho_{R_1T_t}$ and $\rho_{R_0T_t}$ when $\begin{cases} i=j, l=p \\ l=p \\ i=j \end{cases}$, the observation errors are related. Formula (2.44) can be expressed as:

$$Z = h(X) + w \tag{2.67}$$

In the above, Z is a range and measurement:

$$X = [x,y,z]^T, w = \left[w_{\rho_{R_1T_1}}, \cdots, w_{\rho_{R_mT_1}}, \cdots, w_{\rho_{R_1T_n}}, \cdots, w_{\rho_{R_mT_n}} \right]^T$$

The covariance matrix of W is:

$$R = \begin{bmatrix} \delta^2_{\rho_{11}} & \eta_{11}\delta_{\rho_{11}}\delta_{\rho_{21}} & \cdots & \eta_l\delta_{\rho_{11}}\delta_{\rho_{m1}} & \eta_R\delta_{\rho_{11}}\delta_{\rho_{21}} & \cdots & 0 & \cdots & \eta_R\delta_{\rho_{11}}\delta_{\rho_{1n}} & \cdots & 0 \\ \eta_R\delta_{\rho_{m1}}\delta_{\rho_{11}} & \cdots & \eta_R\delta_{\rho_{m1}}\delta_{\rho_{m11}} & \delta^2_{\rho_{nm}} & 0 & \cdots & \eta_R\delta_{\rho_{m1}}\delta_{\rho_{m2}} & \cdots & 0 & \cdots & \eta_R\delta_{\rho_{m1}}\delta_{\rho_{1m}} \\ \eta_R\delta_{\rho_{1n}}\delta_{\rho_{11}} & \cdots & 0 & \cdots & \eta_R\delta_{\rho_{1n}}\delta_{\rho_{1n-1}} & \cdots & 0 & \delta^2_{\rho_{1n}} & \eta_l\delta_{\rho_{1n}}\delta_{\rho_{2n}} & \cdots & \eta_l\delta_{\rho_{1n}}\delta_{\rho_{nm}} \\ 0 & \cdots & \eta_R\delta_{\rho_{1m}}\delta_{\rho_{m1}} & \cdots & 0 & \cdots & \eta_R\delta_{\rho_{1m}}\delta_{\rho_{1m-1}} & \eta_l\delta_{\rho_{1m}}\delta_{\rho_{1n}} & \eta_l\delta_{\rho_{1m}}\delta_{\rho_{1n}} & \cdots & \delta^2_{\rho_{1m}} \end{bmatrix}$$

η_T and η_R are as shown in formula (2.59).

The likelihood functions for the target location X are:

$$P[Z|X] = \frac{1}{\sqrt{|2\pi R|}} \exp\left\{-\frac{1}{2}[Z - h(X)]^T R^{-1}[Z - h(X)]\right\} \qquad (2.68)$$

The Fisher information matrix for the target location estimation is:

$$J = E\left\{[\nabla_X \ln P(Z|X)][\nabla_X \ln P(Z|X)]^T\right\} \qquad (2.69)$$

The final result is shown in formula (2.66).

$$J = [\nabla_X] * [R]^{-1} * [\nabla_X]^T \qquad (2.70)$$

In the above:

$$\nabla_X = -\left[\nabla_X \rho_{R_1 T_1}, \cdots, \nabla_X \rho_{R_m T_1}, \cdots, \nabla_X \rho_{R_1 T_n}, \cdots, \nabla_X \rho_{R_m T_n}\right]$$

$$\nabla_X \rho_{R_l T_i} = \left[\frac{x-x_{R_l}}{r_{R_l}} + \frac{x-x_{T_i}}{r_{T_i}}, \frac{y-y_{R_l}}{r_{R_l}} + \frac{y-y_{T_i}}{r_{T_i}}, \frac{z-z_{R_l}}{r_{R_l}} + \frac{z-z_{T_i}}{r_{T_i}}\right]^T (i = 1, 2, \cdots, n; l = 1, 2 \cdots, m).$$

So, the Cramér–Rao bound P_{CRLB} of the network target location estimation is as follows:

$$P_{\text{CRLB}} = J^{-1} \qquad (2.71)$$

Fig. 2.57 Distribution: centralized central station (one transmitter three receivers)

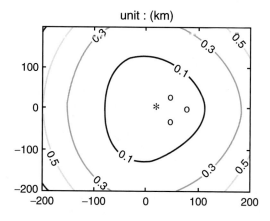

Fig. 2.58 Distribution:
centralized central station (*y*
= 10 km)

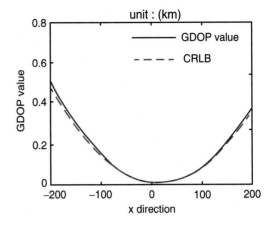

Fig. 2.59 Distribution:
centralized central station
y = 150 km

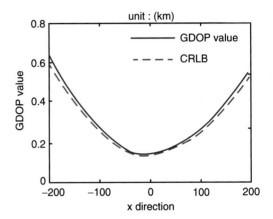

Simulations and Analysis

(a) One transmitter multiple receiver mode

Simulation environment 7 (one transmitter three receivers), transmitting station $T(-10, 0, 0.05)$km, receiving station 1 $R_1(30, 0, 0)$km, receiving station 2 $R_2(0, 30, 0)$km, receiving station 3 $R_3(0, -30, 0)$km. Target height 10 km, the range of target location: x direction ± 200km, y direction ± 200km, range and measurement error standard deviation 3 m, site error 1 m, range and correlation coefficient $\eta = 0.3$; the simulation results are shown in the following figures:

As shown in Figs. 2.57, 2.58, and 2.59, the centralized structure center station locating accuracy of the one transmitter three receiver mode is very high. Taking

Fig. 2.60 Distribution: centralized central station (standard deviation of range and measurement error 10 m)

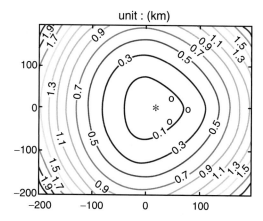

Fig. 2.61 Distribution: centralized central station (standard deviation site error 5 m)

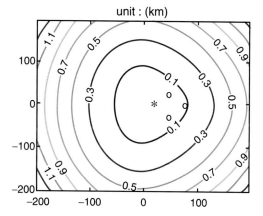

the y direction as an example, in the far range $y = 150$ km or near range $y - 10$ km, the GDOP values approach the Cramér–Rao bound.

When increasing the range and the standard deviation of the measurement error to 10 m, the GDOP distribution located by the centralized structure of the central station is as shown in Fig. 2.60. When increasing the range and standard deviation of the measurement error to 5 m, the GDOP distribution located by the centralized structure of the central station is as shown in Fig. 2.61.

To make a comparison between Figs. 2.60, 2.61, and 2.57, we may find that the increase of the standard deviation of the range and measurement error and site error contributes to the decrease of the central station locating accuracy. The range of GDOP values less than 0.1 km decreases obviously, and the GDOP curve decreases rapidly outwards.

Figure 2.62 shows the distribution of GDOP values of the centralized structure central station location when adding one more receiving station $R_4(-30, 0, 0)$ km on the basis of simulation environment 7. Figure 2.63 shows the distribution of GDOP

Fig. 2.62 Distribution: centralized central station

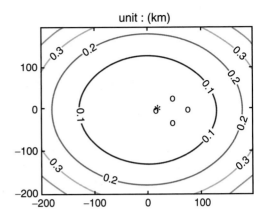

Fig. 2.63 Distribution: centralized central station

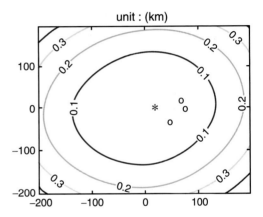

values of the centralized structure central station location when adding yet another receiving station $R_5(21.2, 21.2, 0)$km.

When making a comparison between Figs. 2.62 and 2.57, we may find that, when the number of receiving stations increases to four, the location accuracy of the central station is improved obviously. By making a comparison between Figs. 2.63 and 2.62, we may find that, when the number of receiving stations is more than five, the location accuracy of the central station shows little change. Only the location accuracy on the extension line of the transmitting station and the newly increased receiving station has increased.

Conclusion Under the one transmitter one receiver mode, with the centralized structure location based on range and information, the locating accuracy is very high, approaching the Cramér–Rao bound. Changing the range, measurement error, and site error has an obvious effect on the locating accuracy of the structure. When the number of receiving stations reaches four, the improvement of the locating effect is not obvious, even if the number of receiving stations is further increased.

(b) Multiple transmitter multiple receiver mode

Fig. 2.64 Distribution: centralized central station (three transmitters four receivers)

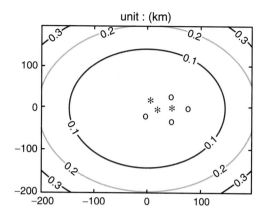

Fig. 2.65 Distribution: centralized central station (four transmitters four receivers)

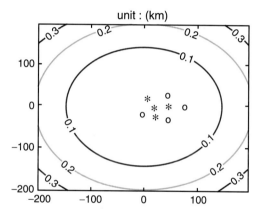

Simulation environment 8 (three transmitters four receivers and four transmitters four receivers). Figure 2.64, based on simulation environment 7, shows the simulation results when adding transmitting station 2, transmitting station 3, and receiving station 4 to the environment. The coordinates of transmitting station 2 are $T_2(10, 0, 0.05)$km, the coordinates of transmitting station 3 are $T_3(0, -10, 0.05)$km, and the coordinates of receiving station 4 are $R_4(-30, 0, 0)$km. Figure 2.65 shows the simulation results obtained following the simulation conditions of Fig. 2.64 after increasing the number of transmitting stations to include $T_4(0, 10, 0.05)$km.

Making a comparison between Figs. 2.64 and 2.62, we can see that the locating accuracy of the central station has improved greatly after increasing the number of transmitting stations by two. The range of GDOP values less than 0.1 km increased obviously. At the same time, the GDOP value decreases outwards.

Making a comparison between Figs. 2.65 and 2.64, we find that the locating accuracy of the four transmitter four receiver mode shows no obvious improvement in comparison with the locating accuracy of the three transmitter four receiver mode.

Conclusion Under the multiple transmitter multiple-receiver mode, the locating accuracy of the centralized structure location based on range and information is very high. With the increase of the numbers of transmitting and receiving stations, the locating accuracy is improved greatly. When the number of transmitting stations is three and the number of receiving stations is four, increasing further the number of receiving stations does not show an obvious effect of improving the locating results.

2.3   Target Location in the Passive Mode

The passive mode can be used to achieve target locating by arrival angle information and the time difference of arrival information. Since broadband characteristics show low location accuracy when using the arrival angle information, while the measurement of the time difference of arrival information can provide high accuracy, therefore, we only discuss and analyze the performance of TDOA locating.

2.3.1   Acquisition of Location Solution

Suppose that target locating under the passive mode is completed by m receiving stations and the central network station on the basis of TDOA, with the central station being used as the locating station. Assume that its location is $R_0(x_0, y_0, z_0)$, the locations of the m receiving stations are $R_i(x_i, y_i, z_i)$ $(i = 1, 2, \cdots, m)$, the target location is $x_T(x, y, z)$, and c is the electromagnetic wave propagation velocity, then the measurement equation is obtained as follows:

$$c\Delta t_i = r_i - r_0 \tag{2.72}$$

where:

$$\begin{aligned} r_0^2 &= (x - x_0)^2 + (y - y_0)^2 + (z - z_0)^2 \\ r_i^2 &= (x - x_i)^2 + (y - y_i)^2 + (z - z_i)^2 \end{aligned} \tag{2.73}$$

Rearranging the equations, we have:

$$(x_0 - x_i)x + (y_0 - y_i)y + (z_0 - z_i)z = k_i + r_0 \cdot \Delta r_i \tag{2.74}$$

where:

$$\Delta r_i = r_i - r_0 = c\Delta t_i$$
$$k_i = \frac{1}{2}\left(\Delta r_i^2 + d_0^2 - d_i^2\right) \quad i = 1, 2, \cdots, m$$
$$d_0^2 = x_0^2 + y_0^2 + z_0^2$$
$$d_i^2 = x_i^2 + y_i^2 + z_i^2 \quad i = 1, 2, \cdots, m$$

In formula (2.70), m equations constitute a set of non-linear equations. An analytical method is applied to solve the equations.

1. Analytical method (one)

First, r_0 is considered as a known quantity, so we can obtain the following matrix expression:

$$AX = F \tag{2.75}$$

where:

$$A = \begin{bmatrix} x_0 - x_1 & y_0 - y_1 & z_0 - z_1 \\ \vdots & \vdots & \vdots \\ x_0 - x_m & y_0 - y_m & z_0 - z_m \end{bmatrix}, X = [x, y, z]^T, F = [k_1 + r_0\Delta r_1, \cdots, k_m + r_0\Delta r_m]^T$$

We select the appropriate location $\text{rank}(A) = 3$, then:

$$X = \left(A^T A\right)^{-1} A^T F \tag{2.76}$$

When x, y, z obtained from the formula (2.72) is substituted into the formula (2.69), we can get the two solutions of r_0, r_{01} and r_{02}. The location ambiguity is obvious and can be omitted by applying a similar practice.

2. Analytical method (two)

When $m \geq 4$, r_0 can be regarded as an unknown quantity. From formula (2.70), we get:

$$\begin{bmatrix} x_0 - x_1 & y_0 - y_1 & z_0 - z_1 & -\Delta r_1 \\ \vdots & \vdots & \vdots & \vdots \\ x_0 - x_m & y_0 - y_m & z_0 - z_m & -\Delta r_m \end{bmatrix} \begin{bmatrix} x \\ y \\ z \\ r_0 \end{bmatrix} = \begin{bmatrix} k_1 \\ \vdots \\ k_m \end{bmatrix} \tag{2.77}$$

Record:

$$BY = g \tag{2.78}$$

Select the appropriate location $\text{rank}(B) = 4$, then according to least squares, we get:

$$\hat{Y} = \left(B^T B\right)^{-1} B^T g \qquad (2.79)$$

The target location $\hat{X}$ can be obtained from $\hat{Y}$, namely, $\hat{X} = HY$ and $H = [I \quad O]$, where I is a unit matrix of order (3×3). O is a zero matrix of order (3×1).

2.3.2 Location Error Analysis

We differentiate both sides of the formula (2.68) to give:

$$cd\Delta t_i = \left(\frac{\partial r_i}{\partial x} - \frac{\partial r_0}{\partial x}\right) dx + \left(\frac{\partial r_i}{\partial y} - \frac{\partial r_0}{\partial y}\right) dy + \left(\frac{\partial r_i}{\partial z} - \frac{\partial r_0}{\partial z}\right) dz + k_i$$

$$+ k_0, \quad i = 1, 2, \cdots, m \qquad (2.80)$$

where:

$$k_i = \frac{\partial r_i}{\partial x_i} dx_i + \frac{\partial r_i}{\partial y_i} dy_i + \frac{\partial r_i}{\partial z_i} dz_i, \quad i = 1, 2, \cdots, m$$

$$k_0 = -\frac{\partial r_0}{\partial x_0} dx_0 - \frac{\partial r_0}{\partial y_0} dy_0 - \frac{\partial r_0}{\partial z_0} dz_0 \qquad (2.81)$$

and:

$$\begin{cases} \dfrac{\partial r_j}{\partial x} = -\dfrac{\partial r_j}{\partial x_j} = \dfrac{x - x_j}{r_j} \\[2mm] \dfrac{\partial r_j}{\partial y} = -\dfrac{\partial r_j}{\partial y_j} = \dfrac{y - y_j}{r_j} \quad j = 0, 1, 2, \cdots, m \\[2mm] \dfrac{\partial r_j}{\partial z} = -\dfrac{\partial r_j}{\partial z_j} = \dfrac{z - z_j}{r_j} \end{cases} \qquad (2.82)$$

Formula (2.76) shows that the target location error is related to the measurement error of the time difference (range difference) parameters and station site error.

The M error equation in formula (2.76) is written in the form of a vector matrix:

$$dV = CdX + dX_s \qquad (2.83)$$

Where:

$$dV = [c \cdot d\Delta t_1, \cdots, c \cdot d\Delta t_m]^T$$

$$dX = [dx, dy, dz]^T$$

$$dX_s = [k_1 + k_0, \cdots, k_m + k_0]^T$$

$$C = \begin{bmatrix} \dfrac{x - x_1}{r_1} - \dfrac{x - x_0}{r_0} & \dfrac{y - y_1}{r_1} - \dfrac{y - y_0}{r_0} & \dfrac{z - z_1}{r_1} - \dfrac{z - z_0}{r_0} \\ \vdots & \vdots & \vdots \\ \dfrac{x - x_m}{r_m} - \dfrac{x - x_0}{r_0} & \dfrac{y - y_m}{r_m} - \dfrac{y - y_0}{r_0} & \dfrac{z - z_m}{r_m} - \dfrac{z - z_0}{r_0} \end{bmatrix} \tag{2.84}$$

By formula (2.79), select the appropriate station site rank$(C) = 3$, and then the pseudo-inverse method can be used to solve the target location error value:

$$dX = (C^T C)^{-1} C^T [dV - dX_s] \tag{2.85}$$

Assume that each measurement error has zero mean and that the white Gaussian noise is not related to each other, the mean square error of the time difference measuring error is $\delta_{\Delta t}^2$, the site of each component between the measurement error and other observation errors and are not related, and with the same mean square error δ_s^2.

Under the above conditions, the covariance matrix of the location error can be deduced as:

$$P_{d\hat{X}} = (C^T C)^{-1} C^T \{ E[dV dV^T] + E[dX_s dX_s^T] \} \left((C^T C)^{-1} C^T \right)^T \tag{2.86}$$

Where:

$$E[dV dV^T] = c^2 \text{diag}[\delta_{\Delta t}^2, \cdots, \delta_{\Delta t}^2]_{m \times m}$$

$$E[dX_s dX_s^T] = \delta_s^2 [I_m + l_m]$$

and where I_m is a unit matrix of order m, and for l_m m order square, its elements are 1.

2.3.3 Cramér–Rao Bound of Target Location Estimation

The measurement error of the network radar system for each receiving station conforms with a zero mean Gaussian distribution. The time difference measurement standard error of $\delta_{\Delta t_i} (i = 1, 2, \cdots, m)$ indicates that, between the i receiving station and the central station TDOA measurement standard error, the measurement errors between the different receiving stations are uncorrelated. Formula (2.68) yields:

$$Z = h(X) + w \tag{2.87}$$

where Z represents the measurement of the time difference:

$$X = [x, y, z]^T \quad w = [w_{\Delta t_1}, \cdots, w_{\Delta t_m}]^T$$

The covariance matrix for w is:

$$R = \text{diag}\left(\delta^2_{\Delta t_1}, \cdots, \delta^2_{\Delta t_m}\right)$$

The likelihood functions for the target location X are:

$$P[Z|X] = \frac{1}{\sqrt{|2\pi R|}} \exp\left\{-\frac{1}{2}[Z - h(X)]^T R^{-1}[Z - h(X)]\right\} \tag{2.88}$$

The Fisher information matrix for the target location estimation is:

$$J = E\left\{\left[\nabla_X \ln P(Z|X)\right]\left[\nabla_X \ln P(Z|X)\right]^T\right\} \tag{2.89}$$

The final result is shown in (2.86):

$$J = [\nabla_X] * [R]^{-1} * [\nabla_X]^T \tag{2.90}$$

where:

$$\nabla_X = -\left[\nabla_X\left(\frac{r_1 - r_0}{c}\right), \cdots, \nabla_X\left(\frac{r_m - r_0}{c}\right)\right]^T$$

In the above formula:

$$\nabla_X\left(\frac{r_i - r_0}{c}\right) = \left[\frac{x - x_i}{r_i \times c} - \frac{x - x_0}{r_0 \times c}, \frac{y - y_i}{r_i \times c} - \frac{y - y_0}{r_0 \times c}, \frac{z - z_i}{r_i \times c} - \frac{z - z_0}{r_0 \times c}\right]^T$$

$r_i(i = 0, 1, \cdots, m)$, as shown in formula (2.69).

So, the Cramér–Rao bound P_{CRLB} of the network center target location estimation is shown as follows:

$$P_{\text{CRLB}} = J^{-1} \tag{2.91}$$

Fig. 2.66 $R_0R_1R_2R_7$ TDOA location

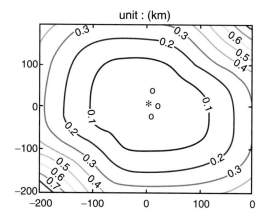

Fig. 2.67 $R_0R_1R_2R_7$ TDOA location ($y = 100\,\text{km}$)

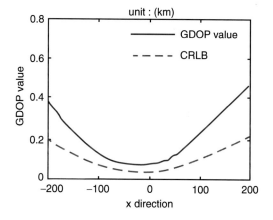

2.3.4 Simulation and Analysis

The location accuracy of the TDOA system is restricted by the relative geometric relationship between the target and the system, and it is especially constrained by the configuration between each station. In recent years, research on the optimal distribution of a multi-station TDOA system abounds. These papers mainly analyze the four station three time difference models, such as a "Y" shaped distribution station, inverted "Y" shaped distribution station, and inverted "T" shaped distribution station. When three-dimensional target locating is applied, for the inverted "T" shaped distribution station when the target is located in the azimuth $60°\sim120°$, we reach the minimum locating error, but the locating error on the line of three receiving stations changes sharply. For the inverted "Y" shaped distribution station, the locating error in any direction changes little, ensuring the same for each direction. So, the inverted "Y" shaped distribution station can improve the locating accuracy when the incoming direction of targets is known. When the incoming

Fig. 2.68 $R_0R_1R_3R_6$ TDOA
location

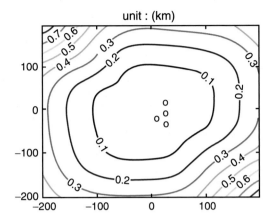

Fig. 2.69 $R_0R_1R_3R_6$ TDOA
location ($y = 100$ km)

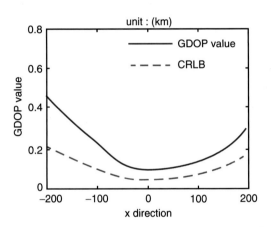

Fig. 2.70 $R_0R_3R_4R_5$ TDOA
location

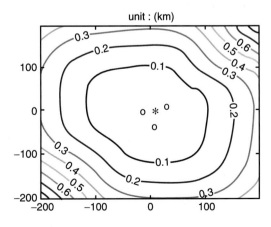

Fig. 2.71 $R_0R_3R_4R_5$ TDOA location ($y = 100\,\text{km}$)

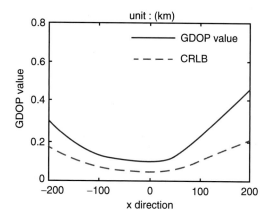

Fig. 2.72 $R_0R_2R_4R_8$ TDOA location

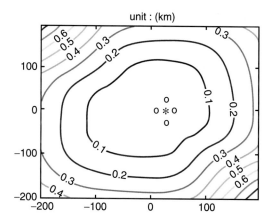

Fig. 2.73 $R_0R_2R_4R_8$ TDOA location ($y = 100\,\text{km}$)

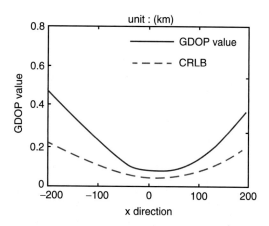

Fig. 2.74 Central station
(fusion of two units)

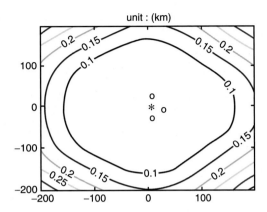

Fig. 2.75 Central (fusion
of four units)

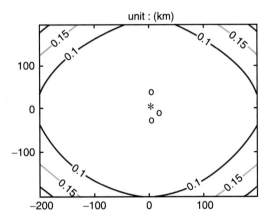

direction of targets is not known or there is the need to search the whole airspace, the inverted "Y" shaped distribution station should be adopted.

Considering the interpretation of the active-passive mode feature and taking the annular distribution station for the condition, the passive mode system can simultaneously constitute multiple independent time difference measuring units, with each group unit approximating an inverted "Y" shape distribution station, and can provide a target location solution to the network center. In the network central station, the fusion of the location solutions can be obtained through the fusion algorithm provided in Sect. 2.2.2.1.1.

Simulation environment 9: The coordinates of the network central station and eight receiving stations are as follows: $R_0 = (0, 0, 0)$km, $R_1 = (30, 0, 0)$km, $R_2 = (0, 30, 0)$km, $R_3 = (0, -30, 0)$km, $R_4 = (-30, 0, 0)$km, $R_5 = (21.2, 21.2, 0)$km, $R_6 = (-21.2, 21.2, 0)$km, $R_7 = (-21.2, -21.2, 0)$km, $R_8 = (21.2, -21.2, 0)$km, site error is 1 m, standard deviation of the measurement errors is 3 ns, target height is 10 km, range of target location: X direction ±200 km,

Fig. 2.76 Increasing the
length of the baseline

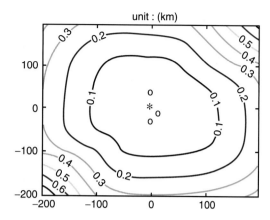

Fig. 2.77 Changing the
time measurement error

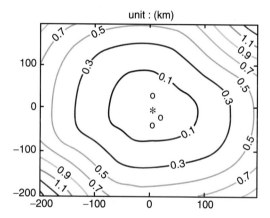

Y direction +200 km, electromagnetic wave propagation velocity $c - 3 \times 10^8$ m/s, $R_0R_1R_2R_7$, $R_0R_1R_3R_6$, $R_0R_2R_4R_8$, and $R_0R_3R_4R_5$ constitute the four approximate inverted "Y" shape distribution locating units.

The simulation results are shown as follows:

A comparison between Figs. 2.66 and 2.70 shows that the TDOA location accuracy of $R_0R_1R_2R_7$ is equal to the TDOA location accuracy of $R_0R_3R_4R_5$.

A comparison between Figs. 2.68 and 2.72 shows that the TDOA location accuracy of $R_0R_1R_3R_6$ is equal to the TDOA location accuracy of $R_0R_2R_4R_8$.

As shown in Figs. 2.67, 2.69, 2.71, and 2.73, taking the Y direction as an example, the GDOP values of the four units approach the Cramér–Rao bound.

As shown in Figs. 2.74 and 2.75, the locating effect of the central station is higher than the effect of each independent unit. When certain location accuracy and operating velocity are both required, two groups of location solutions of the TDOA unit and error of covariance matrix can be sent to the central station.

Taking $R_0R_1R_2R_7$ as the research object, when changing the coordinates of receiving station 7 to $R_7 = (-28.28, -28.28, 0)$km, the simulation results are shown in Fig. 2.76. When changing the standard deviation of the TDOA measurement errors to 1 m, the simulation results are shown in Fig. 2.77.

Making a comparison between Figs. 2.76 and 2.66, increasing the baseline length in one direction leads to the locating accuracy curve improving significantly in the opposite direction, and in the other baseline direction, at the same time as in the opposite direction to the other position of the baseline, the accuracy is almost the same.

Comparing Figs. 2.77 and 2.66, increasing the time difference measurement error standard deviation has a great impact on the overall performance of location searching, and these result in the decrease of locating accuracy.

Conclusion When the system is in the passive TDOA locating mode, due to the independence of multiple time difference localization units and data redundancy, which makes the system have high location precision and strong survival ability, when both certain positioning precision and computing velocity are required, position precision can be sent to the central station in two groups (not quite), as the positioning solution of time unit and an error covariance matrix. If only higher positioning accuracy is required, the positioning solution of the four-group time error and error covariance matrix are sent to the central station for fusion processing. Changing baseline length and the measurement error of the locating accuracy of each independent unit will have a great impact.

2.4 Integration of the Active and Passive Modes for Target Location

When using a surveillance system which integrates active and passive modes, the receiving station can receive the echo signal of the transmitting station, and it can also receive the radioactive signal emitted by the target. At this moment, the system locates the targets using the integrated active mode and passive mode surveillance system. Actually, there are many ways to locate targets. In this section, we study the network center station based on numerous times of arrival.

Taking the one transmitter multiple receiver mode as an example to discuss location principles in the integrated active and passive mode surveillance system in this case, the measurement formulas are as follows:

$$\begin{cases} r_0^2 = (x - x_0)^2 + (y - y_0)^2 + (z - z_0)^2 \\ r_i^2 = (x - x_i)^2 + (y - y_i)^2 + (z - z_i)^2 \quad (i = 1, 2, \cdots, N - 1) \\ \Delta r_i = r_i - r_0 = c \cdot (t_i - t_0) \\ r_T = \sqrt{(x - x_T)^2 + (y - y_T)^2 + (z - z_T)^2} \\ \rho_j = r_T + r_j = c \cdot (t_j - t_T) \quad (j = 0, 1, 2, \cdots, N - 1) \end{cases} \tag{2.92}$$

After completing the simplification, we will obtain formulas as follows:

$$\begin{bmatrix} x_0 - x_1 & y_0 - y_1 & z_0 - z_1 & -\Delta r_1 & 0 \\ \vdots & \vdots & \vdots & \vdots & \vdots \\ x_0 - x_{N-1} & y_0 - y_{N-1} & z_0 - z_{N-1} & -\Delta r_{N-1} & 0 \\ x_T - x_0 & y_T - y_0 & z_T - z_0 & 0 & \rho_0 \\ \vdots & \vdots & \vdots & \vdots & \vdots \\ x_T - x_{N-1} & y_T - y_{N-1} & z_T - z_{N-1} & 0 & \rho_{N-1} \end{bmatrix} \begin{bmatrix} x \\ y \\ z \\ r_0 \\ r_T \end{bmatrix} = \begin{bmatrix} k_1 \\ \vdots \\ k_{N-1} \\ k_0' \\ \vdots \\ k_{N-1}' \end{bmatrix} \tag{2.93}$$

where:

$$k_i = \frac{1}{2} \left[\Delta r_i^2 + (x_0^2 + y_0^2 + z_0^2) - (x_i^2 + y_i^2 + z_i^2) \right], (i = 1, 2, \cdots, N - 1) \tag{2.94}$$

$$k_j' = \frac{1}{2} \left[\rho_j^2 + (x_T^2 + y_T^2 + z_T^2) - (x_j^2 + y_j^2 + z_j^2) \right], (j = 0, 1, 2, \cdots, N - 1) \tag{2.95}$$

Assuming that:

$$\mathbf{A} = \begin{bmatrix} x_0 - x_1 & y_0 - y_1 & z_0 - z_1 & -\Delta r_1 & 0 \\ \vdots & \vdots & \vdots & \vdots & \vdots \\ x_0 - x_{N-1} & y_0 - y_{N-1} & z_0 - z_{N-1} & -\Delta r_{N-1} & 0 \\ x_T - x_0 & y_T - y_0 & z_T - z_0 & 0 & \rho_0 \\ \vdots & \vdots & \vdots & \vdots & \vdots \\ x_T - x_{N-1} & y_T - y_{N-1} & z_T - z_{N-1} & 0 & \rho_{N-1} \end{bmatrix} \tag{2.96}$$

$$\mathbf{X} = \begin{bmatrix} x \\ y \\ z \\ r_0 \\ r_T \end{bmatrix} \tag{2.97}$$

$$\mathbf{F} = \begin{bmatrix} k_1 \\ \vdots \\ k_{N-1} \\ k_0' \\ \vdots \\ k_{N-1}' \end{bmatrix} \tag{2.98}$$

Then the formula (2.89) can be expressed as:

$$\mathbf{AX = F} \tag{2.99}$$

We can select the appropriate site as follows.

Rank $(A) = 5$, with a pseudo-inverse method of solving equations available:

$$\hat{\mathbf{X}} = \left(\mathbf{A}^T\mathbf{A}\right)^{-1}\mathbf{A}^T\mathbf{F} \tag{2.100}$$

Similarly, the location error vector equation can be written in matrix form as follows:

$$\mathbf{dV = CdX + dX_s} \tag{2.101}$$

where:

$$\mathbf{dV} = \begin{bmatrix} d(\Delta r_1) & \cdots & d(\Delta r_{N-1}) & d\rho_0 & \cdots & d\rho_{N-1} \end{bmatrix}^T \tag{2.102}$$

$$\mathbf{C} = \begin{bmatrix} c_{1x} - c_{0x} & c_{1y} - c_{0y} & c_{1z} - c_{0z} \\ \vdots & \vdots & \vdots \\ c_{(N-1)x} - c_{0x} & c_{(N-1)y} - c_{0y} & c_{(N-1)z} - c_{0z} \\ c_{T1} + c_{01} & c_{T2} + c_{02} & c_{T3} + c_{03} \\ \vdots & \vdots & \vdots \\ c_{T1} + c_{(N-1)1} & c_{T2} + c_{(N-1)2} & c_{T3} + c_{(N-1)3} \end{bmatrix} \tag{2.103}$$

$$\mathbf{dX} = \begin{bmatrix} dx & dy & dz \end{bmatrix}^T \tag{2.104}$$

$$\mathbf{dX_s} = \begin{bmatrix} k_0 - k_1 & \cdots & k_0 - k_{N-1} & k_T + k_{R,0} & \cdots & k_T + k_{R,N-1} \end{bmatrix}^T \tag{2.105}$$

The location error estimation value of the target can be solved by the pseudo inverse method, shown as follows:

$$\mathbf{dX} = \left(\mathbf{C}^T\mathbf{C}\right)^{-1}\mathbf{C}^T[\mathbf{dV - dX_s}] \tag{2.106}$$

assuming that:

$$\left(\mathbf{C}^T\mathbf{C}\right)^{-1}\mathbf{C}^T = \mathbf{B} = \begin{bmatrix} b_{ij} \end{bmatrix}_{3 \times n}$$

Similarly, we can know that:

$$\mathbf{P}_{\mathbf{d\hat{X}}} = \mathbf{B}\left\{\mathbf{E}\left[\mathbf{dV} \times \mathbf{dV}^T\right] + \mathbf{E}\left[\mathbf{dX_s} \times \mathbf{dX_s}^T\right]\right\}\mathbf{B}^T \tag{2.107}$$

$$\mathrm{GDOP} = \sqrt{tr\left[\mathbf{P}_{\mathbf{d\hat{X}}}\right]} \tag{2.108}$$

In this case, the Fisher information matrix estimated by the target location can also be expressed as:

Fig. 2.78 Schematic
diagram of network radar
system deployment

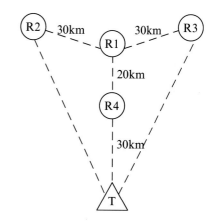

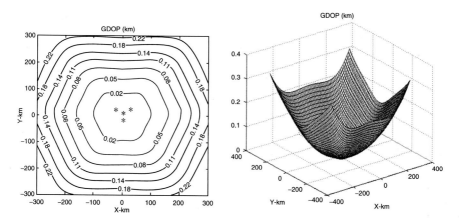

Fig. 2.79 GDOP contour curve two- and three-dimensional maps in passive mode

$$FIM = [\nabla_{\mathbf{X}}]*[\mathbf{R}]^{-1}*[\nabla_{\mathbf{X}}]^T \tag{2.109}$$

Among this formula:

$$\nabla_{\mathbf{X}} = \left[\nabla_{\mathbf{X}}\left(\frac{r_1 - r_0}{c}\right), \cdots, \nabla_{\mathbf{X}}\left(\frac{r_{N-1} - r_0}{c}\right), \nabla_{\mathbf{X}}\left(\frac{r_0 + r_T}{c}\right), \cdots, \nabla_{\mathbf{X}}\left(\frac{r_{N-1} + r_T}{c}\right)\right] \tag{2.110}$$

$$\nabla_{\mathbf{X}}\left(\frac{r_i - r_0}{c}\right) = \left[\frac{x - x_i}{r_i \times c} - \frac{x - x_0}{r_0 \times c}, \frac{y - y_i}{r_i \times c} - \frac{y - y_0}{r_0 \times c}, \frac{z - z_i}{r_i \times c} - \frac{z - z_0}{r_0 \times c}\right]^T \tag{2.111}$$

$$\nabla_{\mathbf{X}}\left(\frac{r_j + r_T}{c}\right) = \left[\frac{x - x_j}{r_j \times c} + \frac{x - x_T}{r_T \times c}, \frac{y - y_j}{r_j \times c} + \frac{y - y_T}{r_T \times c}, \frac{z - z_j}{r_j \times c} + \frac{z - z_T}{r_T \times c}\right]^T \tag{2.112}$$

The CRLB of the target location is:

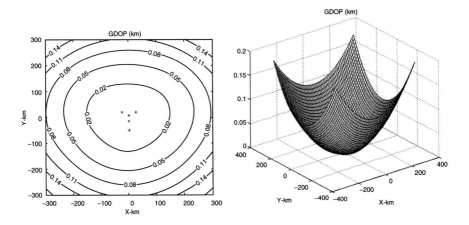

Fig. 2.80 GDOP contour curve two- and three-dimensional maps in active mode

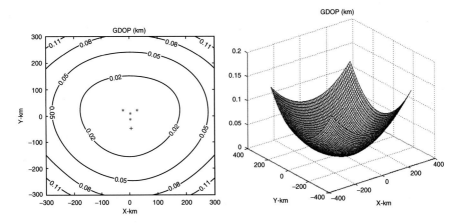

Fig. 2.81 GDOP contour curve two- and three-dimensional maps in integrated active and passive mode

$$\text{CRLB} = (FIM)^{-1} \tag{2.113}$$

Amusing that we use the deployment in Fig. 2.78, the coordinates of each station is: R1 $(0,0)$ km, R2 $\left(-15\sqrt{3}, 15\right)$ km, R3 $\left(15\sqrt{3}, 15\right)$ km, R4 $(0, -20)$ km, and T $(0, -50)$ km, where T is the transmitter station and R is a receiver station.

If the measurement error of the time difference is 1 ns, the correlation coefficient η between the measurement error of each time difference is 0.3, and the location error of each station is 0.1 m. According to the analysis of the location accuracy, we can calculate the network radar system GDOP value of the location error in different working modes. Figures 2.79, 2.80, and 2.81 respectively indicate the

Fig. 2.82 CRLB of target location with the changing of the highest targets in the three modes

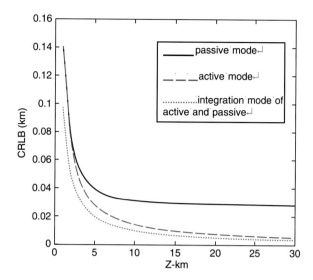

GDOP equivalent map location error on the plane, when working in passive mode, active mode, or integration of the active and passive modes and at a height of 8 km.

According to the simulation results, the most important result is in the development of the integrated active and passive mode, because, at this time, the target location of the system can use the sum of the distance and difference in the distance information at the same time. With more information about target location, the integrated active and passive mode has the most important target location: the location accuracy will be higher, as the active mode has more accurate target location and the passive mode has the worst location accuracy.

Figure 2.82 shows the variation of CRLB of target location with the change of the heights of the target under the three modes, where the horizontal and vertical coordinates of the target are $x = 100$ km, $y = 100$ km.

From Fig. 2.82, we can see that the CRLB value in the active and passive mode is the minimum, showing that this system can achieve optimal location performance.

Chapter 3
Network Radar Countermeasure Systems for Target Recognition

3.1 Introduction

A network radar countermeasure system for target recognition includes two aspects, one referring to the recognition of the radioactive target, namely, electronic target recognition, and the other referring to recognition of the target platform and its associated weapons systems. Some active radar can image the target, such as synthetic aperture or inverse synthetic aperture radar. They can also identify and recognize the target, but, in most cases, it is difficult to identify. Network radar countermeasure systems for target recognition mainly refers to recognition of the radiation source and its loaded platform or associated weapons systems with passive reconnaissance signals. Because the network radar system is compatible with the ability to identify friend or foe, it can also identify friend and foe on the battlefield. At the same time, if active detection of the target speed information is increased, then the identification of the target attributes will become more reliable.

As mentioned above, the main characteristic parameters of target identification are acquired by passive reconnaissance, such as working frequency, pulse repetition frequency, pulse width, frequency modulation, frequency modulation mode, intra-pulse modulation method, frequency variation method, and subtle signal character-istics or fingerprint characteristics, etc.

The process of target recognition begins with the analysis and processing of the signal interceptive parameters to form the description words and characteristic parameters on the radioactive target set, which will be compared with the charac-teristic parameters in the target knowledge database. Finally, the target recognition and identification can be realized based on some kind of identification criterion and confirmation.

Recognition of the target can be completed on both a single station as well as the center station. The identification and evidence based on a single station target forms a logical structure as shown in Fig. 3.1. The target attribute information acquired in the receiving station will be compared with the information of the terminal internal

© National Defense Industry Press, Beijing and Springer-Verlag Berlin Heidelberg 2016   131
Q. Jiang, *Network Radar Countermeasure Systems*,
DOI 10.1007/978-3-662-48471-5_3

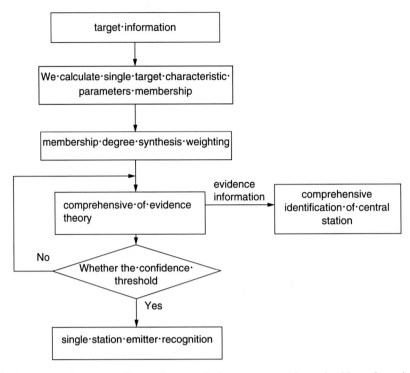

Fig. 3.1 The logic structure diagram between single target recognition and evidence formation

database platform. Firstly, we will calculate the similarity of each parameter of the signal observation samples. The matching calculation for various parameters can determine its similarity. Secondly, we should calculate the membership based on the similarity of each characteristic parameter. Finally, according to the membership of the sample parameter, we should work out a basic probability assignment of evidence, in other words, to acquire a piece of evidence. We should conduct multicycle integration in each of the receiving stations according to the target redundancy in time. With Dempster–Shafe (D–S) evidence combination rules, we should synthesize and redistribute multiple observation samples credibility, and we should obtain evidence from a single station to complete single station target recognition. In the entire identification process, if we can use some characteristic parameters directly, the fine-featured parameters, or identify subtle characteristic parameters of the individual target, the identification process can be greatly simplified. The identification results can also be more reliable.

The network center of the target recognition logic structure is shown in Fig. 3.2. Target identification information gained by the network center station not only includes the target identification information sent by the receiving station, but also includes the target identity information acquired through its own database and battlefield intelligence information through the data chain, etc. The network center station processes the relevant proposition imputed after processing and the existing

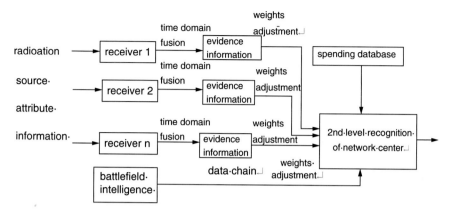

Fig. 3.2 The logic structure of comprehensive target recognition of a network radar countermeasure system

target identification proposition comprehensively. In order to obtain a joint report of the center station network, the network center station adjusts the weights through the source of evidence information under different situations. Comprehensive identification will be carried out in the center station to develop a combined identification report from the network center station.

As mentioned above, emitter individual recognition is an important part of network radar countermeasure system target recognition, and it can identify the specific models, loading platform, and the number of the loading platforms for enemy targets through identifying the individual target platform sources. Finally, it can recognize a radar, an aircraft, or a warship, and then it can obtain information concerning enemy deployment, configuration of weapons and equipment, and their trends.

3.2   Target Recognition with Single Station in Network Radar Countermeasure System

Some factors, such as the multiple working modes of radar target changes and the changing of characteristic parameters, as well as imperfections of the intercepted signal from the radar target, bring about much uncertainty to the task of target recognition. Therefore, the target recognition accuracy and reliability becomes random, fuzzy, and probabilistic. As a result, in the identification process, there is not only one set of target characteristic parameters, but also parameters matched with the identified target set in the database. Therefore, the weighted coefficient of each parameter in the identification of a specific target should be considered differently. At the same time, the tolerance selection of parameters in comparison and the recognition criterion affects the recognition result directly.

3.2.1   The Basic Probability Assignment Calculation of Target Recognition

3.2.1.1   Fuzzy Reasoning

The basic probability assignment function (BPAF) of target recognition is based on fuzzy membership calculation. Fuzziness is a common phenomenon of objective matters. It mainly refers to the "ambiguity" in the transition between objective matters or the uncertainty of the generic boundary or state of a research object. In essence, the fuzzy framework provides a natural way to handle the imprecision source. The definition of the imprecision source is based on the lack of class membership criteria, rather than the lack of the existence of random variables. In the classical or clear set, domain elements in a given set between membership and non-membership changes involves mutation and is easily defined (called "clear"). As for the domain element contained in the fuzzy set, this change is gradual. Changes between different memberships follow the features of uncertainty as well as fuzziness of the fuzzy set's boundary. Therefore, the membership of elements in this collection in the domain can be measured by the function described as uncertain and fuzzy. Because of the diversity and complexity of the traditional fuzzy set method, there are still many imperfections in dealing with the problem, which cannot meet the needs of the specific circumstances. And because of the inherent flexibility of the fuzzy set method, it is hard to form guidelines. In order to overcome the defects of fuzzy pattern recognition, a fuzzy set is generally used in combination with other theories.

Set is the basic concept in mathematics. Fuzzy set is defined as: let U be a universe of discourse and A be a subset; for each $x \in U$, there is a value $\mu_A(x) \in [0, 1]$ that represents the grade of membership of x in A, which means a mapping $\mu_A(x) : U \to [0, 1]$. Then A is called a fuzzy subset of U defined by $\mu_A(x)$, and the function $\mu_A(x)$ is called a membership function. There are different kinds of membership function, such as rectangular function, trapezoidal function, normal distribution, and so on. Choosing a membership function needs to match the characteristics of fuzzy variables, and its parameters should be decided by test.

Three fundamental operations for fuzzy sets are intersection, union, and complement, which are expressed as $\cup, \cap, \neg$. Let A and B be subsets of U, then the membership function of their intersection, union, and complement, respectively, are:

$$A \cup B : \quad \mu_{A \cup B}(x) = \max_{x \in U} \{\mu_A(x), \mu_B(x)\} \tag{3.1}$$

$$A \cap B : \mu_{A \cap B}(x) = \min_{x \in U} \{\mu_A(x), \mu_B(x)\} \tag{3.2}$$

$$\neg A : \quad \mu_{\neg A}(x) = 1 - \mu_A(x) \tag{3.3}$$

The correlation between fuzzy sets will be described by a fuzzy relation. Assuming that the domain $X = \{x_1, x_2, \cdots, x_m\}$ $Y = \{y_1, y_2, \cdots, y_n\}$, then the fuzzy relation R on $X \times Y$ can be expressed with the following membership matrix:

$$
R = \begin{pmatrix} \mu_{11} & \mu_{12} & \cdots & \mu_{1n} \\ \mu_{21} & \mu_{22} & \cdots & \mu_{2n} \\ \cdots & \cdots & \cdots & \cdots \\ \mu_{m1} & \mu_{m2} & \cdots & \mu_{mn} \end{pmatrix} \tag{3.4}
$$

where R is a fuzzy set on $X \times Y$. Assuming that A is a fuzzy relation on $X \times Y$ and B is a fuzzy relation on $Y \times Z$, then the $A \cdot B$ synthesis of fuzzy relation A, B is a fuzzy set on $X \times Z$. Its membership function is:

$$
\mu_{A \bullet B}(x, z) = \bigvee \{\mu_A(x, y) \wedge \mu_B(y, z)\} \tag{3.5}
$$

In the above, $\vee$ indicates $\max(\cdot)$ and $\wedge$ indicates $\min(\cdot)$. Fuzzy synthesis established the method of fuzzy relations on $X \times Z$ produced by fuzzy relations on $X \times Y$ and fuzzy relations on $Y \times Z$.

3.2.1.2 The Fuzzy Membership of Characteristic Parameters

The signals which can be obtained by the receiving station of network radar countermeasure systems through the analysis of parameters and feature extraction include the working frequency, pulse repetition frequency, pulse width, pulse amplitude, frequency modulation, antenna rotations speed, repetition frequency modulation, technology system, and pulse fine features, etc. Assuming that there are N kinds of radar in the gained radar signal template library (recognition framework) $\ddot{U} = \left(\ddot{U}_1, \ddot{U}_2, \cdots \ddot{U}_k, \cdots \ddot{U}_N\right)$, they indicate that there are N kinds of radar target signal or N kinds of working mode of the radar target. Among them, each kind of radar signal $\widetilde{U}_i(j = 1, 2, \cdots N)$ is a joint feature vector containing K characteristic parameters. U_{ij} refers to ith radar and jth parameters. The target radio frequency (RF), pulse repetition interval (PRI), pulse width (PW), and intra-pulse characteristic (IPC) feature data should be a joint feature vector. That is to say, $k = 4$. We also choose the four feature data of the intercepted and received target signals for comparison after feature extraction, the course of which is denoted by $X_m = \{f_m, T_{rm}, P_m, I_m\}$, where m is the measured value. Then, a fuzzy template comparison is made with the fuzzy synthesis evaluation model.

When it comes to the fuzzy comprehensive evaluation of the observed evidence, the parameters' single factor evaluation of the target observation sample multi-feature information must be processed first. In accordance with the type of variable parameters (discrete and continuous analog type), we can use different matching algorithms to determine the membership functions of the characteristic parameters, such as normal and Cushy membership functions. Sometimes, due to the changes of the target signal parameters and some other factors, the signal parameters actually captured by the receiving station vary in a wide range. In order to analyze the target signal and recognition, then according to certain rules, we can determine the appropriate range of each signal parameter, and the tolerance range is called the signal parameters. If the recognized signal X_m by radar really comes from an individual $\bar{U}_i$, about 95 % of the observed value $\{f_m, T_{rm}, P_m, I_m\}$ will fall within U_{ij}, with the corresponding tolerance range σ_{ij} of U_{ij}. So, we can define that various characteristic parameters $\{f_m, T_{rm}, P_m, I_m\}$ belong to the U_{ij} fuzzy membership function, which can be expressed as follows:

1. Membership of the frequency carrier f_m

The frequency carrier is the most basic technical parameter of a radar target. To a large extent, it determines the range of the other technical parameters and also the radar threat types and performance. Therefore, the recognition ratio of the frequency carrier is the highest. For example, for the early warning radar, it requires a long working distance, so we generally choose the VHF and UHF frequency bands. Through the radar equation, we can see that the maximum range of radar is related to the same frequency carrier. Because the maximum range of radar directly determines its purpose, the characteristics of the carrier frequency will affect radar threat types. On a different frequency of the target, the definition based on the carrier frequency membership degree is different. If the frequency type of the $i, i = [1, 2, \cdots N]$th radar targets in the knowledge base is of a fixed frequency, the center frequency is f_i. The difference between the observation frequency and the frequency of the target is $\Delta f_i = |f_m - f_i|$. At the same time, the definition based on the RF membership is as follows:

$$r_f(i) = \begin{cases} 1 & \Delta f_i \leq f_\varepsilon \\ \dfrac{(\Delta f_i - 2f_\varepsilon)^2}{f_\varepsilon^2} & f_\varepsilon < \Delta f_i < 2f_\varepsilon \\ 0 & \Delta f_i \geq 2f_\varepsilon \end{cases} \tag{3.6}$$

In the above, f_ε is the frequency measuring error determined by the system noise and measurement noise. If the frequency type of the ith target is the agility frequency, its center frequency value is f_i, and the agility frequency range is f_A, so the difference value between the observed frequency and the target frequency is $\Delta f_i = |f_m - f_i|$. Here, the membership based on RF is defined as follows:

$$r_f(i) = \begin{cases} 1 & \Delta f_i \leq f_A + f_\varepsilon \\ \dfrac{(\Delta f_i - f_A - 2f_\varepsilon)^2}{f_\varepsilon^2} & f_\varepsilon < \Delta f_i - f_A < 2f_\varepsilon \\ 0 & \Delta f_i \geq f_A + 2f_\varepsilon \end{cases} \tag{3.7}$$

If the frequency of the ith goal type is frequency diversity, taking the frequency of binary as an example, if we set the frequency values as f_{i1}, f_{i2}

respectively, then the difference in frequency is $\Delta f_{ij} = |f_m - f_{ia}|, a = 1, 2$. So the membership degree is defined as follows:

$$r_f(i) = \begin{cases} 1 & \Delta f_{i1} \leq f_\varepsilon, \text{ or } \Delta f_{i2} \leq f_\varepsilon \\ \dfrac{\left(\Delta f_{ij} - 2f_\varepsilon\right)^2}{f_\varepsilon^2} & f_\varepsilon < \Delta f_{ij} < 2f_\varepsilon \\ 0 & \text{else} \end{cases} \tag{3.8}$$

2. Membership of the pulse repetition interval T_{rm}

The PRI refers to the time interval of each target pulse. It determines the radar working distance, and the working distance directly determines the threat type of the radar target. Radar frequency characteristics have obvious constraints on radar target threat characteristics. Through the analysis of sample radar frequency characteristics, we can form the distribution of each type of threat radar frequency data. To different PRIs of the target, the definition of PRI based on the membership degree is different. If the PRI type of the ith goal in the knowledge base is fixed, and the PRI values are T_{ri}, then the difference between observation PRI B and the PRI of the goal is $\Delta T_{ri} = |T_{rm} - T_{ri}|$. Then, the degree of membership based on the PRI is defined as follows:

$$r_{PRI}(i) = \begin{cases} 1 & \Delta T_{ri} \leq T_{re} \\ \dfrac{\left(\Delta T_{ri} - 2T_{re}\right)^2}{T_{re}^2} & T_{re} < \Delta T_{ri} < 2T_{re} \\ 0 & \Delta T_{ri} \geq 2T_{re} \end{cases} \tag{3.9}$$

where T_{re} is the PRI tolerance determined by system noise and measurement noise.

If the PRI type of the ith target is varies heavily in frequency, taking two staggers as an example, assuming that the frame cycle is T_{ri}, the frame cycles of the two staggers are T_{ri1} and T_{ri2}.

The PRI difference is $\Delta T_{rij} = |T_r - T_{ria}|, a = 1, 2$.

At this point, the degrees of membership based on the PRI are defined as follows:

$$r_{PRI}(i) = \begin{cases} 1 & \Delta T_{ri1} \leq T_{re}, \text{ or } \Delta T_{ri2} \leq T_{re}, \text{ or } \Delta T_{ri} \leq T_{re} \\ \dfrac{\left(\Delta T_{ria} - 2T_{re}\right)^2}{T_{re}^2} & T_{re} < \Delta T_{rij} < 2T_{re} \\ \dfrac{\left(\Delta T_{ri} - 2T_{re}\right)^2}{T_{re}^2} & T_{re} < \Delta T_{ri} < 2T_{re} \\ 0 & \text{else} \end{cases}$$

$$\tag{3.10}$$

3. Membership of the pulse width P_m

The pulse width is the duration of the RF signals. The range resolution of the pulse width and radar is related to indicators of speed measuring resolution. Like RF and double frequency, the pulse width can reflect the function and usage of radar to a certain extent. For different kinds of pulse width of the radar, the definition based on PW membership is different. If the center value of the pulse width of the ith target is in the knowledge base, then the difference between the observed target pulse width and the target pulse width is $\Delta P_i = |P_m - P_i|$. At this point, the membership based on PW is defined as follows:

$$r_{PW}(i) = \begin{cases} 1 & \Delta P_i \leq P_\varepsilon \\ \dfrac{(\Delta P_i - 2P_\varepsilon)^2}{P_\varepsilon^2} & P_\varepsilon < \Delta P_i < 2P_\varepsilon \\ 0 & \Delta P_i \geq 2P_\varepsilon \end{cases} \tag{3.11}$$

If the type of pulse width of the ith target is width slippage, assuming that the range of pulse width slippage is P_A, then the membership based on the pulse width is defined as follows:

$$r_{PW}(i) = \begin{cases} 1 & \Delta P_i \leq P_\varepsilon + P_A \\ \dfrac{(\Delta P_i - P_A - 2P_\varepsilon)^2}{P_\varepsilon^2} & P_\varepsilon + P_A < \Delta P_i < 2P_\varepsilon + P_A \\ 0 & \Delta P_i \geq 2P_\varepsilon + P_A \end{cases} \tag{3.12}$$

4. Membership of the intra-pulse fine features I_m

In modern electronic reconnaissance signal processing, sorting and recognizing radar signals reliably also requires analysis of the intra-pulse fine features. Intra-pulse fine features parameters include the signal carrier frequency, bandwidth, amplitude, frequency modulation coefficient, yards wide, and coding rules. For simplicity, we only consider pulse frequency modulation and pulse phase modulation within and without the intra-pulse fine features. Because intra-pulse fine features are discrete variables, the definition of membership based on IPCs is:

$$r_{IPC} = \begin{cases} 1 & \text{When modulation style is the same} \\ 0 & \text{When modulation style is not the same} \end{cases} \tag{3.13}$$

3.2.1.3 Basic Probability Assignment Function of Target Recognition

By using the generalized fuzzy operator, the fuzzy comprehensive evaluation result of the ith radar target obtained in the knowledge base is:

$$m(i) = a_1 r_f(i) \oplus a_2 r_{PRI}(i) \oplus a_3 r_{PW}(i) \oplus a_4 r_{IPC}(i), \\ i = 1, 2 \cdots N \tag{3.14}$$

$$\mu(i) = 1 - \sum_{i=1}^{N} m(i) \tag{3.15}$$

$m(i)$ is the basic probability assignment of the ith radar target in the target database and $\mu(i)$ is the basic probability assignment of an unknown proposition. In the above equations, a_i are the weighting coefficients of the RF, PRI, PW, and the IPC, and $\alpha_1 + \alpha_2 + \alpha_3 + \alpha_4 = 1$, $a_i \geq 0$. In the process of comprehensive evaluation, the weight coefficient a_k directly affects the comprehensive evaluation results. In general, when there is no a priori information on the importance between the target characteristic parameters, we can adopt the method of weighted processing. If we have some prior information, the weights can be determined according to the a priori information.

3.2.2   Target Recognition Based on D–S Evidence Theory

The development of evidence theory, which began in the 1970s, is the extension of classical probability theory. Evidence theory expanded events in probability theory into a proposition and it expanded the collection of events into a set of propositions. It puts forward the basic probability assignment, belief function, and likelihood function (also known as the concept, the establishment of rational function), as well as the relationship between the propositions and sets. Thus, it expanded the problem of proposition uncertainty into the problem of set uncertainty. D–S evidence theory turns reports from receiving stations in evidence, and it constructed the evidence combination model. On the basis of the knowledge and experience of experts in the relevant field, combining multiple evidences through the degree of trust and accumulation of evidence, and with targets constantly changing, we can achieve target recognition with imprecise reasoning methods.

But in the application of D–S evidence theory, the BPAF acquisition is a topic closely related with applications, and it is also the most critical step in practical application because it will directly affect the accuracy and effectiveness of the final decision fusion results. Here, we use the BPAF for the proof method of fuzzy set theory of a fuzzy membership function based on the calculation of the basic probability assignment in the previous section, i.e., the expressions shown in formulas (3.14) and (3.15).

3.2.2.1   D–S Evidence Theory

The recognition framework is a basic concept of the theory of D–S. It consists of a set of incompatible statements. For example, when the recognized target is A_1, A_2....... A_n, then the recognition framework is B and proposition A is any subset of 2^{Ω}. If the proposition is expressed as a vector form, we can use a combination of a vector or single vectors to express some proposition or set of propositions which we are concerned about.

If function $M : 2^\Omega \to [0, 1]$, $M[\varphi] = 0$, and $\sum_{A \subseteq 2^\Omega} M(A) = 1$, then M is the BPAF

on the subset $2\ \Omega$. $M(A)$ is referred to as the basic probability assignment of the proposition and it also indicates the degree of accurate trust to proposition A.

M1 and M2 are two basic probability distribution functions on the recognition framework. With the orthogonal sum rules M $(D) = M1 \oplus M2$, the output of the combination is:

$$M(D) = \begin{cases} \dfrac{1}{1-k} \sum_{D=A \cap B} M_1(A)M_2(B) , & D \neq \phi \\ 0 & , D = \phi \end{cases} \tag{3.16}$$

In the above formula, $k = \sum_{A \cap B = \phi} M_1(A)M_2(B)$, where k is the basic probability

assignment of all contradictory propositions and $M(D)$ is the basic probability assignment of contradictory propositions A and B.

Clearly, formula (3.16) solves the calculation problem of the basic probability assignment of the propositions $<$ and $>$.

If $Bel\, 2^\Omega \to [0, 1]$ and the propositions $B \subseteq A, A \subseteq 2^\Omega$, and $B \subseteq 2^\Omega$ on the basis of the basic belief assignment, then the trust function of A can be defined as:

$$Bel(A) = \sum_B M(B) = \sum_{B \subseteq A} M(B) \tag{3.17}$$

$$-/(\phi) = 0; \ Bel(\Omega) = 1$$

We can verify that the Bel function defined in this way is, indeed, the trust function. Its significance is that the sum of the basic probability assignments of all propositions contained in proposition A (here, it is referred to include proposition C) is defined as the trust function of proposition A, so the Bel function is also known as the lower limit function, indicating all trust to A.

Obviously, formula (3.17) solves the problem of how to calculate the basic reliability distribution of proposition "and" or proposition "or".

If map $Pl : 2^\Omega \to [0, 1]$ and all the $A \in. 2$, $\Omega Pl(A) = 1 - Bel(\overline{A})$ Pl, then the function Pl is the likelihood function. The function Pl is also known as the upper limit function, expressing the degree of faith in A being false and indicating uncertainty to A.

We can prove that, if all $A \in. 2\ \Omega$, $Pl(A) \geq Bel(A)$, then the uncertainty of A can be expressed by $U(A) = Pl(A) - Bel(A)$. $(Bel(A), Pl(A))$ is called the uncertainty interval and it reflects much of the important information of A. The uncertainty of the D–S evidence theory description of A is expressed in Fig. 3.3.

For the BPAF of multiple evidences, with rules of $m(A) = m1 \oplus m2 \oplus \ldots \oplus mn$, then, after combination, the comprehensive probability assignment is as follows:

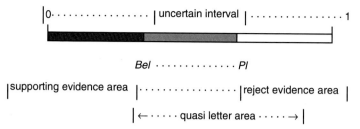

Fig. 3.3 The uncertain description of D–S theory to A

$$M(A) = \begin{cases} 0 & , \quad (A = \phi) \\ \dfrac{1}{1-k}\sum\limits_{\cap A_i = A}\prod\limits_{1 \leq i \leq n} M_i(A_i) = \dfrac{1}{c}\sum\limits_{\cap A_i = A}\prod\limits_{1 \leq i \leq n} M_i(A_i), (A \neq \phi) \end{cases} \quad (3.18)$$

In the above, $i = 1,2,\ldots, n$;

$$k = \sum_{\cap A_i = \phi}\prod_{1 \leq i \leq n} M_i(A_i) \qquad (3.19)$$

$$c = 1 - k = 1 - \sum_{A_i = \phi}\prod_{1 \leq i \leq n} M_i(A_i) = \sum_{\cap A_i \neq \phi}\prod_{1 \leq i \leq n} M_i(A_i) \qquad (3.20)$$

The main ideas for target recognition based on BPAF rules are:

Rule 1: Target classification has the greatest credibility;
Rule 2: The difference in reliability values of the credibility of target classification and other kinds must be greater than a certain threshold ε_1;
Rule 3: The uncertainty interval length must be smaller than a certain threshold ε_2;
Rule 4: The credibility of the target category value must be greater than the uncertainty interval length.

If $\exists A_1, A_2 \subset \Omega$ and:

$$M(A_1) = \max\{M(A_i), A_i \subset \Omega\}$$
$$M(A_2) = \max\{M(A_i), A_i \subset \Omega \text{ and } A_i \neq A_1\}$$

and if:

(a) $M(A_1) - M(A_2) > \varepsilon_1$
(b) $M(U) < \varepsilon_2$
(c) $M(A_1) > M(U)$

then A_1 is the recognition result, in which $\varepsilon_1, \varepsilon_2$ is a preset threshold.

3.2.2.2 The Target Identification Instance Based on D–S Evidence Theory

The results of the target observation of each cycle are fused by the D–S evidence theory. The target recognition process is shown in Fig. 3.4.

Assuming that the main types of airborne radar commonly used include multifunctional attack fire control radar, navigation radar, weather radar, airborne imaging radar, terrain avoidance and terrain following radar, and airborne warning radar, in order to improve the accuracy of target recognition, eliminating recognition caused by random factors and fuzzy recognition errors, each fuzzy recognition results in a proof, and the single station fusion results can be used as evidence sent to the information network center in the receiving station. There is a comprehensive identification based on all kinds of information in the central station. A single receiving station can also make decisions by the combination of evidence information, and it can recognize targets based on a single station.

In order to verify the performance of the algorithm, we simulate four radar targets with a simulator, representing a certain type of airborne early warning radar, another type of airborne early warning radar, airborne radar navigation, and bombing radar, respectively. This target recognition framework is denoted $\Theta = \{A_1, A_2, A_3, A_4\}$. Choosing a simulator to generate an early warning radar, the radar network system receiving station uses three measurement cycles to identify targets.

The first step is to make sure of the BPAF. The BPAF results obtained by fuzzy comprehensive evaluation in the receiving station after each measurement cycle are shown in Table 3.1. Z_1, Z_2, and Z_3 are the observations from different fusion periods.

The four similar radar signals extracted from the radar target feature library form a recognition framework $\Theta = \{A_1, A_2, A_3, A_4\}$, as shown in Table 3.1.

The superposition random measurement error for target A_1 and reconnaissance survey for the signal yields three sample signals, as shown in Table 3.2.

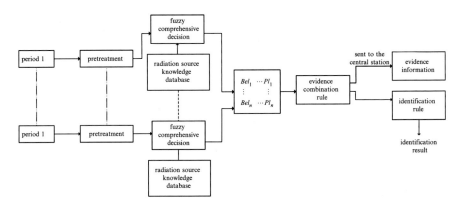

Fig. 3.4 Single station target recognition process diagram

Table 3.1 The feature data of the target

Radar type	RF (MHz)	PRI (μs)	PW (ms)	IPC
A_1	2740	530	0.8	Modulated
A_2	2830	540	0.9	Modulated
A_3	2800	500	1.0	Modulated
A_4	2850	470	1.2	Modulated

Table 3.2 The observed sample data

Observation sample	RF (MHz)	PRI (μs)	PW (ms)	IPC
Z_1	2770	480	0.9	Modulated
Z_2	2730	515	0.85	Modulated
Z_3	2850	545	0.7	Modulated

We use the calculation steps of the fuzzy pattern recognition methods to calculate the characteristic parameters of membership based on the assumption that the characteristic parameters of the weight coefficient is $W = \{0.25, 0.25, 0.25, 0.25\}$. In addition, we should calculate the membership of the feature vector, where M_1 M_2, and M_3 denote the basic probability assignment value function obtained in different cycles based on the membership of the characteristic parameters. The three pieces of evidence acquired during target recognition are shown in Table 3.3.

The BPAF after the combination of the evidence from the three receiving stations M_1, M_2, and M_3 is as shown in Table 3.4

From the calculation results, we can see that the uncertain basic probability assignment decreased to 0.03 following integration. Here, the single receiving station evidence information can be sent to a central station for comprehensive identification. When using rules based on the BPAF, if we choose thresholds $\varepsilon_1 = 0.2$ and $\varepsilon_2 = 0.08$, the final recognition result will be A_1.

3.2.3 Single Station Target Identification

Single station target individual recognition of a network radar system is based on searching for target load emitter fingerprint features for identifying corresponding weapons platform targets. It is relies on the receiving station to complete the target load emitter type recognition, in view of important targets requiring further detailed analysis.

Nowadays, the study of target recognition technology is still in the initial stages. In the 1980s, radar fingerprint analysis was introduced into individual identification. The radar fingerprint features are able to reflect the most essential characteristics of radar parameters and change with the different radar types. There are even subtle differences between the same types of radars. Radiation source fingerprints have the following basic characteristics: (1) universality, radiation source

Table 3.3 BPAF in the different fusion periods

	A_1	A_2	A_3	A_4	U
M_1	0.45	0.25	0.2	0	0.1
M_2	0.31	0.31	0.08	0	0.3
M_3	0.53	0.2	0.07	0	0.2

Table 3.4 BPAF after fusion in different cycles

	A_1	A_2	A_3	A_4	U
$M_{1\times2\times3}$	0.68	0.22	0.07	0	0.03

fingerprints must be characteristic parameters of each radiation source. (2) unique-ness, which means that the fingerprint of each radiation source must be unique, and there are no other identical fingerprints; (3) stability, namely, radiation source fingerprint parameters must be relatively fixed. They should not vary with time, otherwise we cannot obtain accurate measurements. For example, although the radiation source carrier is one of the most important parameters in the identification of the radiation source, due to its variability, it cannot be used as the fingerprint characteristic of the radar signal. Combining the radar emitter recognition devel-opment and previous research results, to complete the emitter recognition, we must solve the following key technologies. Firstly, we should look for radiation source fingerprint characteristic parameters. Second, we must ensure that the radiation source fingerprint characteristic parameters are measurable and can provide com-plete measurements of the fingerprint characteristic parameters. Finally, we should seek a reliable identification method to complete the radar emitter recognition.

3.2.3.1 Selecting the Target Fingerprint Characteristic Parameters

Radar PRI is produced through the procedure in which a crystal oscillator frequency in the target generates a clock signal, and then a clock signal trigger timing signal radar timer generates the corresponding timing signal, which acts later on the radar pulse modulator to generate a pulse time interval of a pulse modulated signal. The PRI process is shown in Fig. 3.5.

According to the radar PRI producing process conclusion, radar PRI jitter is mainly composed of radar timing pulse jitter. That is to say, it is mainly determined by the crystal oscillator in the radar. But there are always some tiny differences among any average frequency of crystal oscillator. According to frequency division theory, frequency division processing will not deteriorate the stability of the output signal. Therefore, the oscillation frequency and frequency stability are unique, and the accurate measurement of the average PRI of the radar will be able to display some subtle differences between radars. Eventually, it is able to achieve recognition of the radar. The frequency stability of the crystal oscillator in the radar is usually located at $10^{-5} \sim 10^{-8}$ orders of magnitude. If the measurement precision PRI of the radar can be higher by one or two orders of magnitude, then it will be fully able to meet the needs of identification of the radar.

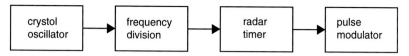

Fig. 3.5 The PRI process

1. Radar target PRI classification.

   There are several kinds of radar signal in pulse mode: fixed PRI, stagger PRI, jittered PRI, stagger PRI jitter, group stagger PRI, etc. The main description of these is shown in Fig. 3.6.

   (1) Fixed PRI

   Ideally, the PRI adjacent to the radar is a fixed deterministic constant. Actually, the unstable vibration of the radar crystal oscillator will cause subtle differences in the PRI_i adjacent to the radar. If the measurement precision of PRI_i is high enough, these subtle differences can be measured.

   (2) Stagger PRI

   The meaning of K stagger PRI is that, in each cycle, a PRI value should be transformed at each K pulse. After multiple cycles, there will be more of the same size fixed set of PRIs. Therefore, it can be dealt with in accordance with the fixed PRI, to some extent.

   (3) Jitter PRI

   The expression for jitter PRI is:

   $$PRI_i = PRI_0 + \delta T_i \tag{3.21}$$

   where PRI_0 is the average of the PRI and δT_i is usually a random sequence within the range of $[-T, T]$, distributed symmetrically. There are many ways to form the PRI jitter, such as sinusoidal modulation, pseudo random sequence modulation, and sampling noise modulation. The ratio $\pm T/PRI_0$ of the jitter value range is T and the center value PRI_0 is known as the maximum amount of jitter. Its typical value is $\pm 1\% \sim \pm 10\%$.

   (4) The jagged jitter PRI and group stagger PRI are combinations on the basis of (1), (2), and (3) above

   The above classification is the most common classification method. Here, from the PRI radar principle, we put forward two methods of reclassifying the radar PRI: one is a constant PRI and the other is changing PRI. The constant PRI is that multiple pulses in the pulse sequence containing the crystal oscillator number corresponding to the same frequency such as PRI and group fixed stagger PRI. Changing PRI is where multiple pulses in the pulse sequence containing the crystal oscillator number correspond to the number of pulse sequences, and the division number is relatively close, such as PRI, PRI and jitter, jitter stagger, and stagger jitter PRI group.

2. Target feature parameter set.

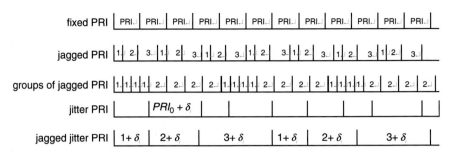

Fig. 3.6 Target PRI classification schemes

There are different characteristic parameters and extraction methods for constant PRI and changing PRI.

Constant PRI:

For a constant PRI pulse sequence, it is generally believed that the PRI's biggest variation is less than 1 % of its average, shown by mathematical expression as:

$$\frac{\max(\Delta PRI)}{\overline{PRI}} \leq 1\% \tag{3.22}$$

For constant PRI, we mainly consider the following characteristic parameters.

(1) Radar PRI estimation

Precise measurement of the radar $\overline{PRI}$ is of great significance, because it can show some subtle changes among the radar. According to $PRI_i = TOA_{i+1} - TOA_i$, the precise measurement of the radar $\overline{PRI}$ depends on the precise measurement of the pulse TOA_i. If the pulse TOA_i is known, then, generally, we can use the arithmetic average method for calculating the radar $\overline{PRI}$. Assuming that the number of pulses is $N + 1$, then the radar $\overline{PRI}$ is expressed as:

$$\overline{PRI} = \frac{1}{N} \sum_{i=1}^{N} PRI_i = \frac{TOA_N - TOA_0}{N} \tag{3.23}$$

This method does not utilize the actual arrival time of the intermediate pulse. A reasonable method is to obtain the pulse interval of the pulse train with the ideal PRI which is the most consistent with the observed pulse TOA. There is a strong universality of this method. First of all we, should assume that the data are measured with a particular functional form. Then, the measured data and function should be fitted, selecting fitting criterion according to the requirements. Assuming that a continuous pulse train is a constant pulse train, in an ideal case, the time interval between adjacent pulses is equal and its value is PRI_0, the number of observed pulses is $N + 1$, and the pulse arrival time is $TOA_0, TOA_1, \cdots, TOA_N$, which satisfies the relationship $TOA_i = i \cdot PRI_0, i = 0, 1, \cdots, N$. In the actual situation, the time of arrival of each pulse is $t_n, n = 0, 1, \cdots, N$, and the error sum of squares of the time of arrival of the ideal pulse and the time of arrival of the actual pulse is:

$$\varepsilon^2 = \sum_{t=0}^{N} (t_i - TOA_i)^2 = \sum_{i=0}^{N} (t_i - i \cdot PRI_0)^2 \tag{3.24}$$

Deriving the above expression, the result is zero, yielding:

$$\frac{\partial \varepsilon^2}{\partial PRI_0} = 2 \sum_{i=0}^{N} (t_i - iPRI_0) \times (-i) = (-2) \sum_{i=0}^{N} it_i + 2 \sum_{i=0}^{N} i^2 PRI_0 \tag{3.25}$$
$$= 0$$

namely:

$$\overline{PRI} = PRI_0 = \left[\sum_{i=0}^{N} it_i \right] \frac{6}{N(N+1)(2N+1)} \tag{3.26}$$

This method makes full use of each pulse's arrival time information. Especially when the measurement of the arrival time information of the pulse is affected by noise, the pulse repetition interval minimum mean square error for the average pulse interval will be lower.

(2) Pulse stability $\delta(PRI)$

From the viewpoint of information theory, modern radar uses a pulse system. The frequency stability of the pulse system and the continuous wave system are different. The pulse width of radar is very narrow, generally of magnitude in the microsecond range. The gap between each pulse is long, generally milliseconds. The pulse repetition frequency is often far less than the upper cut-off frequency of the signal phase noise spectrum. So, under normal circumstances, we cannot obtain the true information of signal source, meaning that it is difficult to use the characterization methods of the frequency stability of the traditional representation for modern radar frequency stability. The radar PRI is generated by a crystal oscillator after frequency division of the radar; therefore, as long as the extraction precision of the PRI is sufficiently high, the radar pulse can objectively reflect the degree of frequency stability of the crystal oscillator in the radar.

The time domain representation of the frequency source stability uses Allan variance, which reflects the difference in the average frequency shift between two adjacent sampling periods. It is expressed as:

$$\sigma(\tau) = \frac{1}{f_0} \sqrt{\sum_{i=1}^{m} \frac{(f_{i+1} - f_i)^2}{2m}} \tag{3.27}$$

where f_0 is the frequency source for a nominal frequency, m is the number of sampling sets, and C and D are, respectively, the ith and the $(i+1)$th time measurement frequency values. According to the parameters which can be obtained by the radar pulse sequence adjacent to the pulse interval PRI_i, the radar pulse stability can be expressed with two kinds of representation.

The first is all continuous sampling, where the sampling number is $m+1$ and the sampling group number is m, as shown in Fig. 3.7.

The pulse stability is expressed as:

$$\delta(PRI) = \frac{1}{\overline{PRF}} \sqrt{\sum_{i=1}^{m} \frac{(PRF_{i+1} - PRF_i)^2}{2m}} \tag{3.28}$$

The second is two adjacent samplings as a group with no clearance within a group, and the clearance time being unlimited. The sample number is $2m$ and the number of sets of samples is m, as shown in Fig. 3.8.

The pulse stability is expressed as:

$$\delta(PRI) = \frac{1}{\overline{PRF}} \sqrt{\sum_{i=1}^{m} \frac{(PRF_{i2} - PRF_{i1})^2}{2m}} \tag{3.29}$$

where $\overline{PRF}$ is the average pulse repetition frequency, and B and C are adjacent pulse repetition frequencies within a group.

From expressions (3.28) and (3.29), it can be seen that the radar pulse stability parameter is closely linked with the sampling time, with the sampling time here being the average pulse repetition interval, so the true meaning of the stability parameter is that the sampling time indicates the stability of the average pulse repetition interval. Different sampling times have different degrees of stability, which is consistent with the meaning of the stability of the crystal oscillator.

(3) Root mean square deviation σ_{PRI}

If we use the root mean square of the crystal oscillator producing radar PRI in one pulse repetition interval of time to express the root mean square $\delta(PRI)$ relative frequency deviation, the root mean square deviation σ_{PRI} of the ideal pulse repetition intervals produced by it can be expressed as:

$$\sigma_{PRI} = \delta(PRI) \cdot \overline{PRI} \tag{3.30}$$

For the radar countermeasure, the acquired parameters are the repetition

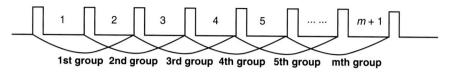

Fig. 3.7 Radar pulse stability characterization, method 1

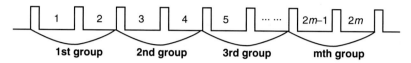

Fig. 3.8 Radar pulse stability characterization, method 2

intervals of each pulse PRI_i, and the $\overline{PRI}$ and $\delta(PRI)$ can be calculated by PRI_i, so the σ_{PRI} can be obtained. If the frequency stability of the reference oscillator used in the process of measuring the pulse repetition interval PRI_i is 1~2 orders of magnitude higher than the frequency stability of the crystal oscillator in the radar, the root mean square deviation of the measurement value of the radar PRI caused by the frequency stability of the reference oscillator is negligible. At this point, σ_{PRI} is able to reflect the properties of the crystal oscillator in the radar.

(4) Range δ_{max}

Range δ_{max} is the difference between the maximum PRI and the minimum PRI in a pulse sequence. This difference can indirectly reflect the performance of the radar PRI producing circuit.

$$\delta_{max} = PRI_{max} - PRI_{min} \tag{3.31}$$

(5) The standard deviation s_{PRI}

The standard deviation can reflect the degree of a discrete pulse sequence, and it is as follows, in accordance with Bessel's formula:

$$v_i = PRI_i - \overline{PRI}$$
$$s = \sqrt{\frac{1}{n-1}\sum_{i=1}^{n}v_i^2} \tag{3.32}$$

The above characteristic parameters are based on the measured value after the second calculation of the radar pulse repetition intervals. Thus, the measuring accuracy of PRI_i can decide whether characteristic parameters can fully reflect the characteristic PRIs of the radar.

Constant PRI must include diverse types. The PRI is significantly gained by different frequencies of the crystal oscillator in a two stagger pulse train; therefore, a two stagger pulse train can be decomposed into two fixed PRI pulses for processing. Similarly, for the n stagger radar pulse signal, we can divide it into fixed PRIs and the characteristic parameters can be extracted from each group of PRIs.

Changing PRI

The PRI types of the pulse sequence transmitted from a radar may contain jumping and shaking. Implementation of these types is produced by controlling the frequency of the crystal oscillator which can generate PRI in the radar. Using programmable control for the divider circuit can easily achieve the change between pulse signal double frequency variation, frequency jitter, and frequency jumps. For the jitter and jumping pulse repetition interval, generally, it is considered that PRI's biggest change is greater than 1 % of its average and less than 30 % of its average, namely:

$$1\% < \frac{\max(\Delta PRI)}{\overline{PRI}} < 30\% \qquad (3.33)$$

For the same type of radar, the crystal oscillator model is established and its frequency is determined. If the frequency can be determined, we can use the correlation method to calculate the clock frequency of the crystal oscillator of the radar. According to the characteristics of the crystal oscillator, there is only one clock frequency.

It is assumed that the clock frequency of the radar crystal oscillator is f_{CLK}, and the repetition interval of the radar pulse is PRI_i. Then, there will inevitably be a relationship type as follows:

$$N_i = f_{CLK} \times PRI_i \qquad (3.34)$$

From the above, A is the number of radar crystal oscillator frequencies.

When the radar uses a number of PRI_i, each PRI_i corresponds to a frequency dividing number N_i, all of which need to meet the above expression. Assuming that the radar oscillator frequency f_{CLK} is fixed, then due to measurement errors ΔPRI, $f_{CLK} \cdot PRI_i$ may not be integers, namely, $\Delta N = f_{CLK} \cdot \Delta PRI$. If $\Delta N > 0.5$, then the N value is uncertain, so the ΔPRI accuracy of PRI must be guaranteed in order to meet the requirements $f_{CLK} \cdot \Delta PRI < 0.5$. In addition, because $N_i = f_{CLK} \cdot PRI_i$, then $m \cdot f_{CLK} \cdot PRI_i$ is also an integer value, which has a multi-value problem. Usually, we assume that the minimum crystal oscillator frequency value which meets the expression is correct, namely, the relative oscillator clock frequency.

The PRI based on the crystal oscillator generator is very stable and any change in the PRIs are usually integer times of a basic interval. Finding out the basic interval is equal to finding out the measured values of the greatest common divisor of all the PRIs. This basic interval can be considered as the lowest frequency of the crystal oscillator which can generate PRI.

Assume that the input target pulse sequence and the extracted pulse time interval are as shown in Table 3.5.

Table 3.5 The interval table of radar pulse repetition

1220.102 μs	1220.104 μs	1214.003 μs	1214.004 μs	1214.006 μs
1226.204 μs	1226.206 μs	1232.305 μs	1281.104 μs	1281.106 μs

The ratio of the range and average of the *PRI* is $\frac{\max(\Delta PRI)}{PRI} = 5.44\%$, which is between 1 % and 30 %. Therefore, we can judge whether it is the shaking type or the jumping type. We should arrange the PRI set from small to large, set and group the tolerances, and calculate the average value. Then, we can take the differences in order to calculate the difference sequence. For the above mentioned data, through the difference processing, we can work out the results, which are shown in Table 3.6.

To set the tolerance for the grouping difference sequence, by considering the actual situation, we can relax restrictions appropriately, considering preliminarily 6.1003 μs as the basic interval. All the PRIs should be divided into basic interval, frequency, and radar clock frequency, as shown in Table 3.7.

From the results, the values of the oscillator frequencies are 199, 200, 201, 202, and 210, and the radar clock frequency is 163.92055 KHz. When it comes to changing the PRI type of the radar, the crystal frequency and crystal oscillator clock frequency are fingerprint feature parameters that should be extracted.

3.2.3.2 The Individual Identification of a Single Station Target

The analysis shows that the radar PRI and other characteristic parameters derived from the radar PRI can be used as a radar fingerprint feature parameter set. The key to this extraction is the high precision measurement of the radar PRI. If the measurement accuracy is not good enough, then the "fingerprint" parameter set extracted according to the PRI will not be able to reflect the individual characteristics of the target.

The high precision time interval measurement method based on interpolating sampling technology includes *ps* order magnitude of measurement accuracy, and the greatest orders of magnitude of the oscillator stability and accuracy are in the range $10^{-5} \sim 10^{-8}$. The radar PRI estimation precision is 1–2 magnitude orders greater than that. The high precision time interval measurement method based on interpolating sampling technology can complete real-time, continuous measurement of pulse intervals. By changing the clock counter digit, we can also change the biggest single corresponding measuring range, which is very suitable for radar PRI measurement. Network radar countermeasure systems can obtain the radar PRI through the high precision time interval measurement method, which is based on interpolating sampling technology. For constant PRI, we should calculate the pulse radar PRI valuations, stability, root mean square deviation, standard deviation, and the range. For changing PRI, we should calculate the frequency multiplication and frequency clock of the radar crystal oscillator. We can then compare the extracted target fingerprint characteristic parameters with the characteristic parameters of the

Table 3.6 PRI average
sequence difference
processing results

PRI collating sequence (μs)	Difference sequence (μs)
1214.004	
1220.103	6.099
1226.205	6.102
1232.305	6.100
1281.105	48.8

Table 3.7 Determination of the frequency and radar clock frequency

PRI ÷ 6.1003	Closest integer	PRI ÷ the closest integer		
193.006934085209	199	6.10051256281407		
193.007261937937	199	6.10052261306533		
193.007589790666	199	6.10053266331658		
200.006884907300	200	6.10051000000000		
200.007212760028	200	6.10052000000000		
201.007163582119	201	6.10051741293532		
201.007491434847	201	6.10052736318408		
202.007278330574	202	6.10051980198020		
210.006720980935	210	6.10049523809524		
210.007048833664	210	6.10050476190476		
	The average interval	6.10051624	Clock frequency	163.92055 KHz

fingerprint database, namely using a radar signal recognition algorithm to complete the individual identification. The individual radar recognition principle is shown in Fig. 3.9.

Individual recognition based on target fingerprint characteristics belongs to the template match problem, namely, to establish range information between unidentified information and known target individual information samples in the target recognition database and to establish the similarity degree between observation information B and each sample of the template base $F(x) = A_i, i = 1, 2, \cdots, n$, so, that, ultimately we can determine the identity of the observed sample information. There are two key problems in radar individual identification: establishing appropriate individual parameter descriptions and selecting suitable matching classification methods.

Here, we set unknown sample $B\left(\overline{PRI}, \delta\left(\overline{PRI}\right), \sigma_{PRI}, s_{PRI}, \delta_{\max}\right)$ as the radar characteristic parameter vector. Radar identification actually classifies the observation samples to the most similar radar individual A_i, which consists of characteristic template target parameter vectors, so it is suitable to use fuzzy closeness as the similarity measure between two feature vector sets. Then, we can compute the degree between the target individual database and observation samples. This degree is set as $N(A_i, B)\big| i = 1, 2, \cdots n$. If $N(A_k, B) = \max\{N(A_i, B)\big| i = 1, 2, \cdots n\}$, then we can conclude that B is mostly close to A_k, so B and A_k are the same individual.

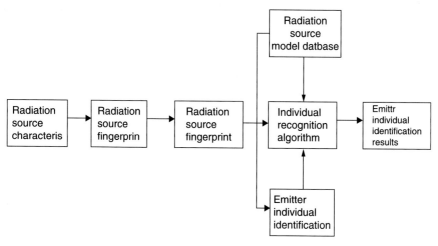

Fig. 3.9 Flow chart of individual target identification

When the near selection method is used for individual identification, there may appear two cases: the first is that object B is not related to individual $A_1, A_2 \cdots A_n$ in each degree, suggesting that B does not recognize $A_1, A_2 \cdots A_n$; the second is that object B is highly related to individual $A_1, A_2 \cdots A_n$ in each degree. In this case, the recognition scope of B can be narrowed. If the pre-condition that B is a member of $A_1, A_2, \cdots, A_n$ is not satisfied, then B may be a new individual in type $F(X)$.

Set threshold $d \in (0, 1]$, order $\alpha = \max\{N(A_i, B) \mid k = 1, 2, \cdots n\}$. If $\alpha < d$, then it cannot be identified. Therefore, B does not belong to any individual of type $F(X)$. Depending on the identification of the radar target type, add B to the radar individual database as new individual information, maintaining the fingerprint individual identification database of the renewing target.

If there is $\alpha \geq d$, and $N(A_1, B) \geq d$, $N(A_2, B) \geq d$, and $N(A_m, B) \geq d$, then B belongs to $A_1 \cap A_2 \cap \cdots A_m$. This kind of situation is not allowed in the individual recognition. In other words, the object is a criterion for multiple individuals. This situation should be processed in accordance with the near selection principle one more time. d Can be determined according to a priori information, for individual radars A_i. After long-term accumulation, there will be many groups of individual characteristic parameter vectors A_{ij}. We calculate the closeness d_{ij} between each vector, and choose the smallest degree $\min(d_{ij})$ as each individual radar distinguishing threshold. In the process of database update, the threshold will be updated accordingly. We should make the distinguishing threshold value of each individual close and choose the minimum value as d. The individual target identification process is shown in Fig. 3.10.

Due to the fact that various characteristics have different weights, we can adopt the method of linear weighting to match the similarity of each feature. According to the weighted value ω_i of each selected characteristic parameter, choose revised Euclid closeness as:

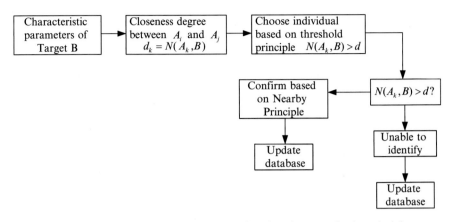

Fig. 3.10 Individual target identification flow chart based on the near selection principle

$$N_E(A,B) = 1 - \sqrt{\sum_{i=1}^{n} \omega_i (A(x_i) - B(x_i))^2} \qquad (3.35)$$

Under the condition without any a priori information, it can only be carried out in accordance with an equally weighted process $\omega_i = 1/n$. We use the principle of characteristic parameters to complete the standard deviation coefficient. In the long-term accumulation process of the radar individual database, the same type of different radar individual records grouped together many characteristic parameters. Take constant PRI as an example. We use $\left(\overline{PRI}, \delta(\overline{PRI}), \sigma_{PRI}, s_{PRI}, \delta_{\max}\right)$ as its characteristic parameters. Except for $\delta(\overline{PRI})$, all the parameters have the same units, and they are calculated from the same radar pulse sequence. Therefore, the standard deviation characteristic parameter set can reflect the influence degree in the recognition.

Each of the sets of characteristic parameters should have the standard deviation λ_i calculated in turn, The lower the λ_i, the smaller the degree of separation of their corresponding characteristic parameters, and the smaller the differences between individuals.

The smaller its influence on identification, then the smaller the corresponding weight ω_i. Because the pulse stability parameter $\delta(\overline{PRI})$ does not have any units, the weight is given according to the size of the average. Here, it is 0.2.

ω_i is calculated according to the formula:

$$\omega_i = \frac{\lambda_i}{\sum_{i=1}^{n} (\lambda_i)} \times 0.8 \qquad (3.36)$$

In the sequence of B and A_i, the unit of each characteristic value is different. It must be dimensionless because it is processed before the calculation of the closing degree. Setting the reference sequence $\alpha = [\alpha_1, \alpha_2, \cdots, \alpha_n]$, each unit of the sequence is the same as the corresponding characteristic parameters sequence. The sequences B and A_i are calculated as:

$$B = B.{*}\alpha$$
$$A_i = A_i.{*}\alpha \qquad\qquad (3.37)$$

In actual operation, let $\alpha = 1./B$, then we can complete the dimensionless processing of sequences B and A_i.

With the increase of radar individual data records, the individual database needs to be updated accordingly. Shown in Fig. 3.11 is the radar individual recognition database hierarchy diagram. Every radar individual corresponds to multiple records. In the database update process, first of all, we complete the updating of the weights ω_i, according to formula (3.36), and take the mean value of all the records of the individual A_i, obtain the characteristic parameter vector of A_i, and calculate the discrimination threshold d_i of the corresponding individual.

Assume that a certain type of radar has three individuals, A1, A2, and A3, with each individual having five records, as shown in Table 3.8.

According to the individual records, calculate the characteristic parameters of the corresponding weight coefficients, as shown in Table 3.9.

Calculate the computing degree between the individual records according to the weight coefficients. In order to perform the calculation conveniently, we can calculate the computing degree between each record and their average records. The results are shown as d_i in Table 3.8. Select a minimum of 0.9783, 0.9810, and 0.9891 as the thresholds.

Input the same type of radar characteristic value sequence B = (1030.02256, 7.48 e-5, 7.48, 1030.02256, 510.453), and, in turn, calculate the computing degree of their gravity center of individuals A1, A2, and A3. The results are 0.9815, 0.8716, and 0.7454, according to the principle of threshold and near selection; therefore, the radar will be judged to be individual A1.

3.3   Network Center Comprehensive Target Recognition

Target identification information received by the network center includes the time fusion target recognition information, which is offered by each station in all measurement cycles. The central station obtains target identification information through supporting database resources (such as mobile rules, the target platform attributes, the goal target type, and number of statistical results). Also, it includes all non-realistic information, such as: battlefield intelligence, target occurrence probability, and the target type, quantity, and tasks, which are received by the data link type. Each level of target identification information will also be used as evidence of the center of a comprehensive target, using the fusion rule to comprehensively determine identification of the target.

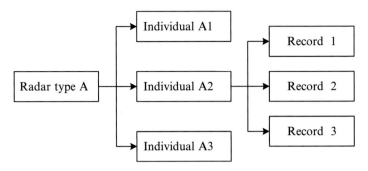

Fig. 3.11 Radar individual recognition database hierarchy

Table 3.8 Radar individual characteristic parameters

		$\overline{PRI}(\mu s)$	$\delta(\overline{PRI})$	$\sigma_{PRI}(ns)$	$s_{PRI}(ns)$	$\delta_{max}(ns)$	d_i
Individual A1	Record 1	1030.02165	7.69e-5	73.209	90.008	508.584	0.9965
	Record 2	1030.02182	7.43e-5	76.531	83.987	507.378	0.9783
	Record 3	1030.02174	7.54e-5	77.664	83.868	507.376	0.9891
	Record 4	1030.02190	7.76e-5	73.929	83.976	507.745	0.9901
	Record 5	1030.02186	7.85e-5	80.857	83.906	507.834	0.9820
	A1 average record	1030.021794	7.654e-5	78.8380	83.9490	507.7834	0.9783
Individual A2	Record 1	1030.02354	8.66e-5	83.200	100.008	510.564	0.9936
	Record 2	1030.02376	8.47e-5	87.243	96.967	517.037	0.9810
	Record 3	1030.02378	8.52e-5	87.758	97.878	516.396	0.9874
	Record 4	1030.02345	8.78e-5	90.436	93.996	510.785	0.9877
	Record 5	1030.02380	8.82e-5	90.848	93.986	517.730	0.9849
	A2 average record	1030.023666	8.65e-5	83.0970	98.9670	514.5024	0.9810
Individual A3	Record 1	1030.02535	6.67e-5	68.703	80.006	507.184	0.9944
	Record 2	1030.02556	6.57e-5	67.673	73.997	506.478	0.9946
	Record 3	1030.02576	6.55e-5	67.467	73.968	506.379	0.9924
	Record 4	1030.02548	6.58e-5	67.776	73.866	506.758	0.9957
	Record 5	1030.02518	6.72e-5	63.218	73.967	506.889	0.9891
	A3 average record	1030.025466	6.62e-5	68.1674	73.9608	506.7376	0.9891

Table 3.9 Results of the characteristic parameter identification weights

	$\overline{PRI}$	σ_{PRI}	s_{PRI}	δ_{max}	$\delta(\overline{PRI})$
Standard deviation (ns)	1.55943	8.94325	8.07229	4.04177	–
Weight	0.0552	0.3163	0,2855	0.1430	0.2

3.3.1 Evidence Weighted Processing of Central Station

In the synthesis rules of D–S evidence, we usually assume that the importance of each piece of evidence is the same, namely, each evidence contributed the same amount to the overall result. However, the network countermeasure system of each receiving station provides different evidence capabilities in different environments. Each station has different detection performances for different targets. In a complex environment, as a result of the existence of deception and jamming, the importance of evidence which is provided by the receiving station in a fusion system is different. Other information with different attributes carry different amounts of information, such as the greater differences in the target category values, which are given by particular attribute information, greater resolving power, larger amount of information contained, and more important role in the fusion and final decision. These are not reflected in the D–S evidence synthesis formula. In this case, directly using Dempster synthesis integration may deteriorate the performance of the whole system.

So, we introduce the concept of evidence weight, and convert the evidence weights to make their values equal. Then, we use the Dempster evidence synthesis method to identify the multi-stations target attributes in airspace.

Assume that the system's trust in evidence E is α, and then set $1 - \alpha$ as the discount rate.

Definition 1 Assuming that $m : 2^{\Omega} \rightarrow [0, 1]$ is a basic probability distribution function, then when $0 \le \alpha \le 1$, let $m^{\alpha} : 2^{\Omega} \rightarrow [0, 1]$, suiting:

$$
\begin{aligned}
&m^{\alpha}[\varphi] = 0; \\
&m^{\alpha}[\Omega] = \alpha m(\Omega) + (1 - \alpha); \\
&m^{\alpha}[A] = \alpha m(A); \\
&\forall A \in 2^{\Omega}, A \ne \Omega, \text{and } A \ne \varphi
\end{aligned}
\tag{3.38}
$$

Then, m^{α} is a basic probability distribution function.

In definition 1, element Ω in the power set 2^{Ω} is said to be a collection of all the basic propositions in the identification framework. It does not provide any information, only equivalents of unknown propositions. So, $m(\Omega)$ stands for unknown propositions of the basic probability distribution function.

After evidence E, of which the discount rate of E is $1 - \alpha$, converts the basic probability distribution function m into m^{α} by definition 1, the credibility α to evidence E is directly reflected in the proposition of the basic probability distribution function in the system.

Definition 2 Set the relative weights as $\omega_1, \omega_2, \cdots \omega_r$ for evidence $E_1, E_2, \cdots E_r$, respectively. If $\omega_m = \max\{\omega_1, \omega_2, \cdots \omega_r\}$, then ω_m can be taken as the key evidence and the rest are non-critical evidence. The weight of each evidence relative to the key evidence is $\alpha_i = \omega_i/\omega_m, i = 1, 2, \cdots r$. Obviously, the right value of the key evidence is 1.

Evidence weight is an important measure in all evidence, which reflects the degree of contribution to the final synthesis results. When using Dempster synthesis rules, the relative weight of each evidence should be equal. So, the first thing to consider before synthesis is the relative weight of the evidence. On the evidence of the basic probability distribution function, we use definition 1 to convert the weights of the various evidence of the basic probability distribution function, and make evidence weight of each transformed evidence be equal.

Set the recognition framework as $\Omega = \{A_1, A_2 ....A_n\}$, then each evidence E is independent from one other. $m_j(B_i)$ is the basic probability distribution function for proposition $B \subset 2^\Omega$, which is provided by the evidence. For evidence E_i, according to evidence weight α_i, which is relative to the key evidence, it can be used for converting the basic probability distribution function as follows:

$$m_j'(B_i) = \alpha_i m_j(B_i);$$
$$m_j'(\Omega) = \alpha_i m_j(\Omega) + (1 - \alpha_i) \tag{3.39}$$

After such a conversion, all the evidence weights of the evidence can be considered as being equal. After transforming, the relative importance of each evidence can be reflected directly by the basic probability distribution function. After transforming the basic credibility of each evidence to all propositions, we can use the Dempster synthesis rules for synthesis.

3.3.2   Recognition Framework Adjustment of Central Station

The target recognition information obtained through the process of integrated identification information in the network center includes real-time information of each station, target determination result obtained by database resources data processing, and target information obtained by the data chain, etc. Existing problems among each evidence that is sent to the fusion center include the lack of evidence scope, the ambiguity of reporting problems, the contradiction between various evidences, etc. If we integrate these evidences directly, on one hand, that may increase the uncertainty of the conclusion, so there may even be conflicting conclusions. On the other hand, in the process of calculation, there will appear some jamming factors whose probability value is small, which may lengthen the reasoning chain and the recognition framework becomes more complicated. Thus, it will bring about a very large amount of calculation in the process of system operation.

When evidence theory is considered as the uncertain reasoning model, rules can be easily added or deleted. In order to improve the accuracy of the computing speed and conclusions, we add a group of information filters to the fusion system to modify the uncertainty of the system and get rid of the jamming factor.

The network center will carry out algorithm isolation on the predicted target so as to not appear in the recognition frame, which is obtained by intelligence. At the same time, in the process of integration, we set the rejection threshold, getting rid of some small probability values jamming, filtering, and screening small probability events, and assign Ω to the basic probability of small probability targets, to narrow the recognition framework and speed up the convergence.

Assume that the center of intelligence shows that there are k types of uncertain target appearing in the target recognition framework Ω to obtain a target set X of k types of target. There are l types of goal, which is impossible to judge and the target set is Y. p types of target do not appear, so these p target types are set for Z, and $k + l + p = n$, as per the original target recognition framework. Take the rejection threshold as σ^j. If the revised basic probability is assigned $m_j'(A_i) \leq \sigma_j$, this target is isolated in the fusion process. We can see that the isolated target number should be greater than or equal to p. If q types of target are filtered through algorithm isolation, reorder them after normalization processing.

$$\overline{m}_i(A_i) = \frac{m_j'(A_i)}{\sum\limits_{i=1}^{n-q} m_j'(A_i) + m_j'(\Omega)} \tag{3.40}$$

Identify the framework $\Omega = (A_1, A_2, \cdots A_{n-q})$. After adjustment:

$$\overline{m}_i(\Omega) = 1 - \sum\limits_{i=1}^{n-q} m_j'(A_i) \tag{3.41}$$

Then, we use Dempster combination (as mentioned earlier in this section) for target identification to make the judgment results of the target.

3.3.3 Integrated Identification Example of Central Station

To verify the above method, we consider more than one source of evidence; for example, use a simulator to simulate six kinds of radar target. They are: a certain type of airborne early warning radar, airborne navigation radar, and bombing targeting radar. Choosing the target recognition framework for $\Theta = \{A_1, A_2, A_3, A_4, A_5, A_6\}$, the evidence sources include the basic probability assignment $m_{ESM1}(\cdot)$ and $m_{ESM2}(\cdot)$ obtained by two stations through fuzzy comprehensive evaluation, and a basic probability assignment $m_{data}(\cdot)$ obtained through database resource processing. The information obtained through the data link shows some uncertainty of A_1, A_2, A_3, A_4 could not be determined, and A_5, A_6 do not appear. $m_l(\theta) = 0.1$. Because terminal 2 is affected by the environment and jamming, the relative weights of terminal 1 and terminal 2, database and the data link were 0.2, 0.1, 0.3, and 0.4, respectively, which is given by the expert system.

Table 3.10 Basic probability assignment table

	Relative weights	A_1	A_2	A_3	A_4	A_5	A_6	U
$m_{ESM1}(\cdot)$	0.2	0.34	0.28	0.11	0.16	0	0.05	0.06
$m_{ESM2}(\cdot)$	0.1	0.42	0.24	0.12	0.08	0.04	0	0.10
$m_{data}(\cdot)$	0.3	0.4	0.3	0.05	0.15	0	0.05	0.05
$m_l(\cdot)$	0.4	0.5	0.4	0	0	$\otimes$	$\otimes$	0.10

Table 3.11 Modified basic probability assignment table

	Evidence weight	A_1	A_2	A_3	A_4	A_5	A_6	U
$m'_{ESM1}(\cdot)$	0.5	0.17	0.14	0.055	0.08	0	0.025	0.53
$m'_{ESM2}(\cdot)$	0.25	0.105	0.06	0.03	0.02	0.01	0	0.775
$m'_{data}(\cdot)$	0.75	0.3	0.225	0.0375	0.1125	0	0.0375	0.2875
$m'_l(\cdot)$	1	0.5	0.4	0	0	$\otimes$	$\otimes$	0.10

Table 3.12 Fixed basic probability assignment table

	A_1	A_2	A_3	A_4	U
$\overline{m}_{ESM1}(\cdot)$	0.1746	0.1436	0.056	0.082	0.5438
$\overline{m}_{ESM2}(\cdot)$	0.106	0.06	0.03	0.02	0.784
$\overline{m}_{data}(\cdot)$	0.3117	0.2238	0.0390	0.1169	0.3086
$\overline{m}_l(\cdot)$	0.5	0.4	0	0	0.1

The original combination of the basic probability assignment is shown in Table 3.10.

It can be seen that the relative weights of the basic probability distribution function is the largest, which is obtained by the data link. We select it as the key evidence, take its value as 1, and adjust the relative weights of the basic probability assignment obtained by the other channels to make them equal. Again, according to formula (3.39), the basic probability assignment after the evidence is weighted is shown in Table 3.11.

Using the truncation type filtering fusion processing model, set the rejection threshold as $\sigma_j = 0.015$. By assuming that A_5 and A_6 were algorithmically isolated, we can get the correct basic probability assignment, as shown in Table 3.12.

If we take the rejection threshold as $\sigma_j = 0.02$, then A_4, A_5, and A_6 were algorithmically isolated, so we can obtain the modified basic probability assignment, which is shown in Table 3.13.

Table 3.14 shows the Dempster combination results of using no truncation type filtering fusion processing, using the rejection threshold $\sigma_j = 0.015$, and using the rejection threshold $\sigma_j = 0.02$.

We can see from Table 3.14 that the results of the three identification methods are consistent. The network center used weighted fusion processing, fully considering the effectiveness of evidence provided by each station. Through target filtering and selection of small probability events, the target recognition framework is reduced, and the average recognition time is also cut down. In particular, under

Table 3.13 Adjustment of
the basic probability
assignment table

	A_1	A_2	A_3	U
$\overline{m}_{\text{ESM1}}(\cdot)$	0.1899	0.1564	0.0614	0.5923
$\overline{m}_{\text{ESM2}}(\cdot)$	0.1082	0.0619	0.0309	0.799
$\overline{m}_{\text{data}}(\cdot)$	0.3529	0.2647	0.0441	0.3383
$\overline{m}_{I}(\cdot)$	0.5	0.4	0	0.1

Table 3.14 Three conditions of the fusion results

	A_1	A_2	A_3	A_4	A_5	A_6	U
First	0.5672	0.3631	0.0097	0.0209	0	0.0061	0.0327
Second	0.5745	0.3609	0.0098	0.0211	$\otimes$	$\otimes$	0.0337
Third	0.5830	0.3735	0.01	$\otimes$	$\otimes$	$\otimes$	0.0336

the condition of increasing target recognition units, expanding the target search scope, and increasing the computational cost, comprehensive target recognition accuracy and real-time performance of the network radar countermeasure system is greatly enhanced. Thus, in a way, it improves the performance of the center target recognition system.

Target recognition is an important function of radar countermeasure systems, and is also an important basis of situation and threat assessment. According to their own characteristics, network radar countermeasure systems are divided into two levels: the attribute information recognition based on a single receiving station and target fusion identification based on the center station. Through the combined application of D–S evidence theory and fuzzy reasoning, we effectively solved the allocation problem of the basic probability distribution function, formed each individual terminal's evidence and target recognition, and then analyzed the reason why radar PRI can be used as a fingerprint characteristic parameter. On that basis, the radar fingerprint characteristic parameter and its calculation methods are studied and the radar individual identification is analyzed and verified. Aiming at the features of different types of evidence information obtained by the network center, we put forward the concept of evidence weight to modify a variety evidences obtained in different ways. Through the truncation of small probability events, we reduced the target recognition framework and improved recognition instantaneity.

Chapter 4
Target Tracking of Network Radar Countermeasure Systems

4.1 Introduction

Target tracking of a network radar countermeasure system involves two problems; one is a tracking problem based on a single target and the other is a tracking problem based on simultaneous multiple targets. In the active multiple transmitter one receiver mode and passive mode, focusing on the tracking problem of a single receiving station and a central station with regards to a single target, for target tracking, we can use a centralized processing filter or a distributed processing filter. A centralized processing structure means that all the data are transmitted to the terminal measurement of the center for centralized processing and fusion. A distributed processing structure means that each station does some pre-treatment first, and then sends the intermediate results to the fusion center. In dealing with the centralized structure, the central station can make use of the original measurement data of all stations without any information loss. Thus, we can achieve the optimal fusion result, but the structure needs a wideband data transmission link to transmit the raw data, and that requires a center with great processing power. Distributed structure processing has a lower requirement for digital links and processing power. It has higher real-time performance, but during processing, it may bring about information loss.

4.2 Target Motion Model

The moving target model is one of the basic elements of target tracking, and is also a critical and difficult issue. For maneuvering targets, the ideal model is very difficult to establish, because, in most cases, a priori knowledge of the target maneuvering is poorly understood. The maneuvers are anthropogenic and often controlled by humans, which makes them difficult to accurately describe with

© National Defense Industry Press, Beijing and Springer-Verlag Berlin Heidelberg 2016   163
Q. Jiang, *Network Radar Countermeasure Systems*,
DOI 10.1007/978-3-662-48471-5_4

mathematical expressions. Only in a variety of conditions can it be described with an approximate method. When the moving target model is being built, the general principle is that the established model needs to conform to the actual maneuvering target state, but it also needs to be easy in terms of mathematical treatment. Generally speaking, the target motion model of state variables can choose the location and speed of the object. The target motion model of state variables can choose the location of the object as well as its speed. The state variables increase the amount of estimating calculations required. Therefore, under the condition of meeting the tracking performance, we often use a simple mathematical model. The following is a simple analysis of the acceleration model, Singer model, and maneuvering target model, which are commonly used in the movement target tracking field under a uniform model. A comparison of their advantages and disadvantages is presented.

4.2.1   Uniform Motion Model

Assume that the state vector $X = [x, \dot{x}]^T$ is in the ideal state $\ddot{x}(t) = 0$. Consider that, as a result of the actual existence of noise disturbance, constant speed cannot be absolutely accurate. Assuming the slight change described as zero mean white noise:

$$\ddot{x}(t) = w(t) \tag{4.1}$$

then, apparently, the smaller the variance σ^2 of w(t), the closer the constant speed value. At this time, the target of the continuous time state equation is:

$$\begin{bmatrix} \dot{x} \\ \ddot{x} \end{bmatrix} = \begin{bmatrix} 0 & 1 \\ 0 & 0 \end{bmatrix} \begin{bmatrix} x \\ \dot{x} \end{bmatrix} + \begin{bmatrix} 0 \\ 1 \end{bmatrix} w(t) \tag{4.2}$$

Assume that the sampling period is T, then the discrete time state equation will be:

$$X_{k+1} = F_{CV}X_k + W_k \tag{4.3}$$

where:

$$F_{CV} = \begin{bmatrix} 1 & T \\ 0 & 1 \end{bmatrix} \tag{4.4}$$

The covariance matrix of W_k is:

$$Q_{CV} = E[W_k W_k^T] = \begin{bmatrix} T^3/3 & T^2/2 \\ T^2/2 & T \end{bmatrix} \sigma^2 \tag{4.5}$$

In the uniform motion model, any unnecessarily introduced components (such as acceleration) can only lead to poor performance tracking. For the uniform motion in a straight line or the approximate uniform motion in a straight line, this model can achieve very high tracking precision.

In actual application, the uniform motion model can also be used to approximately describe the maneuvering target state. The difference between the uniform motion model and the non-motor vehicle state of uniform motion model is that the state noise variance is often bigger. This is because, when the goals are in a state of motor, and the acceleration is larger, a larger state noise needs to be introduced. Using the uniform motion model to represent the target maneuvering model is one of the simplest ways commonly used in place of smaller target maneuvering or random states. But when the target is in a state of strong maneuvering, using the uniform motion model can cause a larger error, so, at this moment, we need to fully consider the maneuvering state of the target.

4.2.2 Uniformly Accelerated Motion Model

Assume the state vector $X = [x, \dot{x}, \ddot{x}]^T$, then disturbance of the acceleration in the ideal condition is $\dddot{x}(t) = 0$. Consider that, as a result of the existence of noise disturbance, constant acceleration cannot be absolutely accurate. Assuming a slight change in the zero mean white noise, then:

$$\dddot{x}(t) = w(t) \tag{4.6}$$

Apparently, the smaller the variance σ^2 of $w(t)$, the nearer to constant the acceleration value will be. So, the continuous time state equation for the target is:

$$\begin{bmatrix} \dot{x} \\ \ddot{x} \\ \dddot{x} \end{bmatrix} = \begin{bmatrix} 0 & 1 & 0 \\ 0 & 0 & 1 \\ 0 & 0 & 0 \end{bmatrix} \begin{bmatrix} x \\ \dot{x} \\ \ddot{x} \end{bmatrix} + \begin{bmatrix} 0 \\ 0 \\ 1 \end{bmatrix} w(t) \tag{4.7}$$

The set of sampling period is T and the discrete time state equation is:

$$X_{k+1} = F_{CA} X_k + W_k \tag{4.8}$$

where:

$$F_{CA} = \begin{bmatrix} 1 & T & T^2/2 \\ 0 & 1 & T \\ 0 & 0 & 1 \end{bmatrix} \tag{4.9}$$

The covariance matrix of W_k is:

$$Q_{CA} = E[W_k W_k^T] = \begin{bmatrix} T^5/20 & T^4/8 & T^3/6 \\ T^4/8 & T^3/3 & T^2/2 \\ T^3/6 & T^2/2 & T \end{bmatrix} \sigma^2 \tag{4.10}$$

For acceleration in the straight line motion or approximately uniformly accelerated motion, the model can achieve very high tracking precision.

4.2.3   Singer Model

In 1970, Singer. R.A put forward the zero-mean and time-correlated maneuvering acceleration model. The model assumes that the biggest target maneuvering acceleration is A_{max}, with probability P_{max}, non-motor vehicle probability is P_0 (i.e., the acceleration is zero), acceleration is in the range $(-A_{max}, A_{max})$, approximately obeying a uniform distribution, and the probability density function is:

$$P(a) = [\delta(a - A_{max}) + \delta(a + A_{max})]P_{max} + \delta(a)P_0$$
$$+ [u(a + A_{max}) - u(a - A_{max})]\frac{1 - P_0 - 2P_{max}}{2A_{max}} \tag{4.11}$$

where $u(\cdot)$ is the unit step function and $\delta(\cdot)$ is the impulse response function.

Assume that the maneuvering acceleration time correlation function is of the exponential decay form, namely:

$$R_a(\tau) = E[a(t)a(t + \tau)]$$
$$= \sigma_a^2 e^{-\alpha|\tau|} \quad (\alpha \geq 0) \tag{4.12}$$

In the above formula, σ_a^2 α is in the range $(t, t + \tau)$, determining the parameters of the target maneuvering characteristics. α is the reciprocal of the motor time constant, namely, the maneuvering frequency, and σ_a^2 is the maneuvering acceleration variance. According to the assumption of acceleration probability distribution, it can be calculated as:

$$\sigma_a^2 = \frac{A_{max}^2}{3}[1 + 4P_{max} - P_0] \tag{4.13}$$

Using the time correlation function $R_a(\tau)$, after maneuvering, the acceleration is a (t), obtained by Wiener–Kolmogorov filtering. It can be expressed as white noise, namely:

$$\dot{a}(t) = -\alpha a(t) + w(t) \tag{4.14}$$

In this formula, $w(t)$ is the white noise, which has zero mean and the Gaussian variance is $2\alpha\sigma_a^2$. Thus, we can obtain the target maneuvering model expressed by white noise, for the one-dimensional case, using the Singer statistical model as:

$$\begin{bmatrix} \dot{x} \\ \ddot{x} \\ \dddot{x} \end{bmatrix} = \begin{bmatrix} 0 & 1 & 0 \\ 0 & 0 & 1 \\ 0 & 0 & -\alpha \end{bmatrix} \begin{bmatrix} x \\ \dot{x} \\ \ddot{x} \end{bmatrix} + \begin{bmatrix} 0 \\ 0 \\ 1 \end{bmatrix} w(t) \tag{4.15}$$

For the one-dimensional case, based on the Singer statistical model, the target state equation is:

$$X_{k+1} = F_{Singer}X_k + W_k \tag{4.16}$$

In the above equation, the state transition matrix F_{Singer} is:

$$F_{Singer} = \begin{bmatrix} 1 & T & \frac{1}{\alpha^2}(-1 + \alpha T + e^{-\alpha T}) \\ 0 & 1 & \frac{1}{\alpha}(1 - e^{-\alpha T}) \\ 0 & 0 & e^{-\alpha T} \end{bmatrix} \tag{4.17}$$

The discrete time process W_k with noise covariance is:

$$Q_{Singer} = E[W_k W_k^T] = 2\alpha\sigma_a^2 \begin{bmatrix} q_{11} & q_{12} & q_{13} \\ q_{12} & q_{22} & q_{23} \\ q_{13} & q_{23} & q_{33} \end{bmatrix} \tag{4.18}$$

where:

$$q_{11} = \frac{1}{2\alpha^5}\left[1 - e^{-2\alpha T} + 2\alpha T + \frac{2\alpha^3 T^3}{3} - 2\alpha^2 T^2 - 4\alpha T e^{-\alpha T}\right]$$

$$q_{12} = \frac{1}{2\alpha^4}\left[e^{-2\alpha T} + 1 - 2e^{-\alpha T} + 2\alpha T e^{-\alpha T} - 2\alpha T + \alpha^2 T^2\right]$$

$$q_{13} = \frac{1}{2\alpha^3}\left[1 - e^{-2\alpha T} - 2\alpha T e^{-\alpha T}\right]$$

$$q_{22} = \frac{1}{2\alpha^3}\left[4e^{-\alpha T} - 3 - e^{-2\alpha T} + 2\alpha T\right]$$

$$q_{23} = \frac{1}{2\alpha^2}\left[e^{-2\alpha T} + 1 - 2e^{-\alpha T}\right]$$

$$q_{33} = \frac{1}{2\alpha}\left[1 - e^{-2\alpha T}\right]$$

As $\alpha \to 0$, $F_{\text{Singer}} = \begin{bmatrix} 1 & T & T^2/2 \\ 0 & 1 & T \\ 0 & 0 & 1 \end{bmatrix}$, as in the uniformly accelerated motion model.

And as $\alpha \to \infty$, $F_{\text{Singer}} = \begin{bmatrix} 1 & T & 0 \\ 0 & 1 & 0 \\ 0 & 0 & 0 \end{bmatrix}$, as the model of uniform linear motion.

When α changes on the real number line continuously, the Singer model is corresponds to different motion states, while the target varies from uniform motion to uniformly accelerated motion. The Singer model is essentially an a priori model adaptive tracking model, because it does not make use of the online information for target maneuvering. One of the main shortcomings of the Singer model is its balance, namely, at any time, the target acceleration has a zero mean value. In fact, in the absence of online information for target maneuvering, it is the best model to use.

4.2.4 The Turning Model of Maneuvering Targets

The target turning motion is often referred to as a coordinated turn. Its movement characteristic is the angular velocity, where the speed of the target remains the same, but its direction is changing. However, because of jamming, the actual turning motion is not a coordinated turn, but a variable noise can be used to approximately compensate for the model error. Set a state vector $X = [x, \dot{x}, y, \dot{y}]^T$, then the turning movement model in two-dimensional space can be described by the following equation:

$$X_{k+1} = F_{CT}X_k + G_{CT}W_k \tag{4.19}$$

where:

$$F_{CT} = \begin{bmatrix} 1 & \dfrac{\sin(wT)}{w} & 0 & \dfrac{1-\cos(wT)}{w} \\ 0 & \cos(wT) & 0 & -\sin(wT) \\ 0 & \dfrac{1-\cos(wT)}{w} & 1 & \dfrac{\sin(wT)}{w} \\ 0 & \sin(wT) & 0 & \cos(wT) \end{bmatrix}, G_{CT} = \begin{bmatrix} {T^2}/{2} & 0 \\ T & 0 \\ 0 & {T^2}/{2} \\ 0 & T \end{bmatrix}$$

In this formula, w is the target turning rate, $W(k)$ is the state disturbance noise vector with mean value zero and variance $Q(k)$. Relative to the other motor model, this description is a special kind of motion.

4.3   Tracking Filtering Algorithm

The filtering problem is the most basic element in a target tracking system. The purpose of filtering is to estimate the present motion state of the target, including its position, speed, and acceleration. After a long period of study, scholars put forward many filtering methods for tracking moving targets under different conditions.

4.3.1   Information Filter

The information filter is an inverse matrix for calculating the covariance matrix on forecasting and updating recursively. In general, when the system state dimension m is much larger than the measuring vector dimension n, there are some advantages in using the standard Kalman filter, because it is an inverse matrix of dimensions $m \times m$. However, when the measuring vector dimension n is much larger than the system state dimension m, using an information filter can greatly reduce the amount of calculation required, as its inverse matrix has dimensions $n \times n$. Information filters are widely used in information fusion.

Consider a system as follows:

$$\begin{cases} X_{k+1} = F_k X_k + \Gamma_k W_k \\ Z_k = H_k X_k + V_k \end{cases} \tag{4.20}$$

Assume that the Kalman filter exists, and that it is reversible so as to transfer all states to a matrix F_k, then all covariance matrices are also reversible; X_k is the state vector of system at the kth moment; $W_k \sim N(0, Q_k)$ is the evolution process of noise, which is an independent process; Γ_k is a noise matrix; Z_k is the measuring vector for the system state at the kth moment; H_k is a measurement matrix; and $V_k \sim N(0, R_k)$ is the measured noise, which is also an independent process. It is mutually independent to W_k and both are independent their own initial states. The covariance matrix and the matrix of the Kalman filter gain can be calculated according to the information filter method:

1. Step forecast information for the matrix:

$$P_{k/k-1}^{-1} = \left(A_{k-1}^{-1} + \Gamma_{k-1}Q_{k-1}\Gamma_{k-1}^T\right)^{-1} \tag{4.21}$$

where $A_{k-1}^{-1} = F_{k-1}P_{k-1/k-1}F_{k-1}^T$.

2. The matrix of the filtering information:

$$P_{k/k}^{-1} = P_{k/k-1}^{-1} + H_k^T R_k^{-1} H_k \tag{4.22}$$

3. Kalman information for the matrix expression of the gain matrix:

$$K_k = P_{k/k}H_k^T R_k^{-1} \tag{4.23}$$

4.3.2   Non-linear Filtering Algorithm

Assume that the state equation of the system is:

$$X_{k+1} = F_k X_k + \Gamma_k W_k \tag{4.24}$$

In this formula, the concrete value of the target motion state according to the type of target motion may be obtained (refer to Sect. 4.2).

The measurement equation is as follows:

$$Z_k = h(X_k) + V(k) \tag{4.25}$$

In formulas (4.24) and (4.25), except for measurement equation in formula (4.25), which is of the non-linear form, the other expressions are the same as in formula (4.20).

4.3.2.1   EKF Filtering Algorithm

Based on the target motion equation of state (4.24) and the observation (4.25) for measuring the parameters, the EKF filtering process is as follows.

One-step prediction:

$$\hat{X}_{k/k-1} = F_k \hat{X}_{k-1/k-1} \tag{4.26a}$$

Prediction error covariance:

$$P_{k/k-1} = F_k P_{k-1/k-1} F_k^T + \Gamma_k Q_k \Gamma_k^T \tag{4.26b}$$

Kalman gain:

$$K_k = P_{k/k-1} H^T \left[H P_{k/k-1} H^T + R_k \right]^{-1} \tag{4.26c}$$

The filter:

$$\hat{X}_{k/k} = \hat{X}_{k/k-1} + K_k \left[Z_k - H \hat{X}_{k/k-1} \right] \tag{4.26d}$$

Filter covariance:

$$P_{k/k} = [I - K_k H] P_{k/k-1} \tag{4.26e}$$

where:

$$H = \left. \frac{\partial h(\cdot)}{\partial X} \right|_{X = \hat{X}_{k/k-1}} \tag{4.27}$$

4.3.2.2   UKF Filtering Algorithm

Although the application of EKF filtering to non-linear system state estimation has been widely approved and adopted, there is an obvious defect. For the sake of the spread of calculating the estimation error covariance of the EKF, we use Taylor linearization for the dynamic model in the current estimation state, and Taylor linearization for the measurement model in the one-step prediction state. Because the Taylor series expansion method has shortcomings in that the overall properties of a function (its average) are replaced by local characteristics (its derivatives), the existence of noise makes it worse. To improve the effect of the filtering problem of a non-linear model, Julier proposed the unscented Kalman filter (UKF)method. When using this method in the treatment of the state equation, the unscented transformation is carried out first (U transformation), and the state variables are used after U transformation for filter estimation, thus reducing the estimation error.

Based on the target motion state (4.24) and the observation (4.25) for measuring the parameters, the UKF filter process can be divided into the three following steps:

The first step is to produce a rough estimate of the target initial state $\hat{X}_{k-1|k-1}$ and covariance matrix $P_{k-1|k-1}$. We then calculate the target prediction step $\hat{X}_{k|k-1}$ with the UT method, as well as the prediction error covariance matrix $P_{k|k-1}$.

(a)   While calculating the σ of sampling point $\xi^{(i)}_{k-1|k-1}$ of the target initial state $\hat{X}_{k-1|k-1}$ $(i = 0, 1, \cdots L)$, we usually take $L = 2n$ and set n as the dimensions of the target state variable X_k:

$$
\begin{cases}
\xi^{(0)}_{k-1|k-1} = \hat{X}_{k-1|k-1} \\
\xi^{(i)}_{k-1|k-1} = \hat{X}_{k-1|k-1} + \left(\sqrt{(n+\lambda)P_{k-1|k-1}} \right)_i \quad i = 1, 2, \cdots, n \\
\xi^{(i)}_{k-1|k-1} = \hat{X}_{k-1|k-1} - \left(\sqrt{(n+\lambda)P_{k-1|k-1}} \right)_i \quad i = n+1, n+2, \cdots, 2n
\end{cases}
\tag{4.28}
$$

(b)   Calculate σ for sample point $\xi^{(i)}_{k-1|k-1}$ $(i = 0, 1, \cdots 2n)$ by the spread of the state equation, to obtain the target state step prediction $\hat{X}_{k|k-1}$ and the prediction error covariance matrix $P_{k|k-1}$:

$$
\begin{cases}
\xi^{(t)}_k = F_k \xi^{(t)}_{k-1|k-1} \\
\hat{X}_{k|k-1} = \sum_{i=0}^{2n} \omega_i^{(m)} \xi^{(i)}_k \\
P_{k|k-1} = \sum_{i=0}^{2n} \omega_i^{(c)} \left(\xi^{(i)}_k - \hat{X}_{k|k-1} \right) \left(\xi^{(i)}_k - \hat{X}_{k|k-1} \right)^T + \Gamma_k Q_{k-1} \Gamma_k^T
\end{cases}
\tag{4.29}
$$

In the above, $\lambda = \alpha^2(n+\kappa) - n$ decides the point spread degree of point σ. We usually take a small positive number (e.g., 0.01). Constant κ is the second proportion parameter, which we usually we as 0. $\left(\sqrt{(n+\lambda)P_{k-1|k-1}} \right)_i$ is the first ith column of the square root of the matrix. $\omega_i^m (i = 0, 1, \cdots, 2n)$ is the weight coefficient while calculating the first-order statistical properties and $\omega_i^{(c)} (i = 0, 1, \cdots, 2n)$ is the weight coefficient while calculating the second-order statistical properties. Usually, $\omega_0^m = \lambda/(n+\lambda)$, $\omega_0^{(c)} = \lambda/(n+\lambda) + (1 - \alpha^2 + \beta)$, $\omega_i^m = \omega_i^{(c)} = 0.5/(n+\lambda)$, $i = 1, 2, \cdots, 2n$, β describe the distribution information of the target state (in a Gaussian distribution, the optimal value is 2).

In the second step, we use UT to solve the spread of σ at the sampling point of one-step prediction $\hat{X}_{k|k-1}, P_{k|k-1}$ through the measurement equation, to obtain the step predicted value of the measurement as $\hat{Z}_{k|k-1}$. Thus, the new rate of covariance S_k and filtering gain matrix K_k are calculated.

(a) Calculate the sample point of $\hat{X}_{k|k-1}, P_{k|k-1}$:

$$\begin{cases} \xi_k^{(0)} = \hat{X}_{k|k-1} \\ \xi_k^{(i)} = \hat{X}_{k|k-1} + \left(\sqrt{(n+\lambda)P_{k|k-1}}\right)_i & i = 1, 2, \cdots, n \\ \xi_k^{(i)} = \hat{X}_{k|k-1} - \left(\sqrt{(n+\lambda)P_{k|k-1}}\right)_i & i = n+1, n+2, \cdots, 2n \end{cases} \quad (4.30)$$

(b) Calculate one-step ahead prediction $\hat{Z}_{k|k-1}$ of the output measurement, the new covariance S_k, and filtering gain matrix K_k:

$$\begin{cases} \zeta_k^{(i)} = h\left(\xi_k^{(i)}\right) & i = 0, 1, \cdots, 2n \\ \hat{Z}_{k|k-1} = \sum_{i=0}^{2n} \omega_i^{(m)} \zeta_k^{(i)} \\ S_k = \mathrm{cov}\left[\hat{Z}_{k|k-1}\right] = \sum_{i=0}^{2n} \omega_i^{(c)} \left(\zeta_k^{(i)} - \hat{Z}_{k|k-1}\right)\left(\zeta_k^{(i)} - \hat{Z}_{k|k-1}\right)^T + R_k \\ \mathrm{cov}\left[\tilde{X}_{k|k-1}, \tilde{Z}_{k|k-1}\right] = \sum_{i=0}^{2n} \omega_i^{(c)} \left(\xi_k^{(i)} - \hat{X}_{k|k-1}\right)\left(\zeta_k^{(i)} - \hat{Z}_{k|k-1}\right)^T \\ K_k = \left[\tilde{X}_{k|k-1}, \tilde{Z}_{k|k-1}\right] S_k^{-1} \end{cases} \quad (4.31)$$

In the third step, after obtaining the new measurement Z_k, process the filter updates:

$$\begin{cases} \tilde{X}_{k|k} = \tilde{X}_{k|k-1} + K_k\left(Z_k - \hat{Z}_{k|k-1}\right) \\ P_{k|k} = P_{k|k-1} - K_k S_k^{-1} K_k^I \end{cases} \quad (4.32)$$

The EKF is the minimum mean square error (MMSE) estimation of a first-order Taylor expansion based on the evolution of the system state function and measurement function. It actually uses the local linearization method of the current state to approximate the system state evolution equation. But, most of the time, the local linearization method may lead to a not very ideal approximation effect, or even filter divergence. Both the UKF and EKF perform Gaussian approximation on the system state posterior distribution, but their difference mainly lies in the fact that the EKF is the first-order Taylor expansion depending on a non-linear function only. But when the system is highly non-linear and Taylor expansion of a high order has a large impact on the system, the EKF method may lead to filter divergence, and there is a requirement to calculate the derivative of a non-linear function. The UKF method still uses a Gaussian distribution to characterize the system state variables, but it uses the actual non-linear model with a minimum set of sample points to approximate the system state distribution function. The sample set can accurately capture the mean value and variance of the system state. According to the actual non-linear model evolution, in theory, we can capture the posterior mean value of any non-linear function and variance of the second-order items.

4.3.3   Adaptive Filtering Algorithm

Generally speaking, based on several filtering methods described earlier, it is difficult for a filter to have a good tracking performance under various target motions. Therefore, many scholars have been devoted to the study of various adaptive filtering algorithms and obtained many important results. A typical adaptive filtering algorithm is the interacting multiple model algorithm.

4.3.3.1   Interacting Multiple Model Algorithm

Blom and Bar-Shalom put forward a kind of interacting multiple model (IMM) algorithm with Markov switching coefficients algorithm. IMM estimation uses a reasonable assumption management technology, generally considered to be one of the most effective hybrid estimation schemes, which has been successfully applied to many practical problems, and has gradually become the mainstream direction of research in this field. The IMM estimator is one of the best known state estimators with single scanning. While using the IMM algorithm for calculation of the target state estimation, considering each model filter likely to be the current effective filter system model, the initial conditions of each filter is based on the filtering results of the synthetic conditions model of previous model. While realizing the IMM algorithm, the selection of the model set is very important, and sometimes even more important than the algorithm itself. Theoretically, if we can establish all possible motion models, namely, if the model set is complete, then we can obtain the optimal estimates of the target motion state. However, multi-model filter estimates are based on Bayesian inference, which asks each event to be independent and complete, so meticulous modeling may also reduce the differences in models and affect the relevancy of the model, thus damaging the basis of the optimality algorithm. Usually, several typical motion models can be used to approximately describe the movement characteristics of the target. Assume that the target motion model is set for $\{M_1, M_2, \cdots, M_r\}$, then the first j motion model is:

$$X_{k+1}^j = F_j X_k^j + \Gamma_j W_k^j \tag{4.33}$$

In this formula: X_k^j is the target state vector; F_j is the state transition matrix; Γ_j is the input control matrix; and W_k^j is the Gaussian white noise sequence of system to meet $E\left(W_k^j\right) = 0$, $E\left(W_k^j W_k^j T\right) = Q_k^j$.

The jumping rules of the model are subject to a Markov chain process with the following transition probability:

$$P\{M(k) = M_j/M(k-1) = M_i\} = p_{ij} \tag{4.34}$$

In this formula, p_{ij} is the transition probability of the model when the system transfers from the i to j model based on the Markov chain. The IMM algorithm process can be summarized as the following four steps.

The first step is to determine the mixture initial state of the various filtering models. Assuming that the first j model is valid under the condition of the current time, we can obtain the initial conditions of the filters matching the first j model by mixing the conditions estimation of all filters in the previous time:

$$\hat{x}^{0j}_{k-1/k-1} = \sum_{i=1}^{r} \hat{x}^{i}_{k-1/k-1} \mu^{i/j}_{k-1/k-1} \tag{4.35}$$

$$P^{0j}_{k-1/k-1} = \sum_{i=1}^{r} \mu^{i/j}_{k-1/k-1} \left\{ P^{i}_{k-1/k-1} + \left[\hat{x}^{i}_{k-1/k-1} - \hat{x}^{0j}_{k-1/k-1}\right] \left[\hat{x}^{i}_{k-1/k-1} - \hat{x}^{0j}_{k-1/k-1}\right]^{T} \right\} \tag{4.36}$$

In the above:

$$\mu^{i/j}_{k-1/k-1} = P\{M_{k-1} = M_i/M_k = M_j, Z^{k-1}\}$$
$$= \frac{1}{\overline{C}_j} P\{M_k = M_j/M_{k-1} = M_i, Z^{k-1}\} P\{M_{k-1} = M_i/Z^{k-1})$$
$$= \frac{1}{\overline{C}_j} p_{ij} \mu^{i}_{k-1}$$

where $\overline{C}_j$ is the normalized constant and $\overline{C}_j = \sum_{i=1}^{r} p_{ij} \mu^{i}_{k-1}$ states the probability of the first $k-1$ moment model.

The second step is filtering. Corresponding to the first k moment model M_j, according to the initial conditions of the first step and the current measurement parameters Z_k, set it as the input of each filter of the first k moment and produce a new model estimation $\hat{x}^j_{k/k}$ and $p^j_{k/k}$. The filtering process is as follows:

To predict $\hat{x}^j_{k/k-1} = F_j \hat{x}^{0j}_{k-1/k-1}$

Prediction error covariance: $p^j_{k/k-1} = F_j p^{0j}_{k-1/k-1} F_j^T + \Gamma_j Q_k^j \Gamma_j^T$

Gain: $K_k^j = p^j_{k/k-1} H_j^T \left[H_j p^j_{k/k-1} H_j^T + R_k^j\right]^{-1}$

Filter: $\hat{x}^{j}_{k/k} = \hat{x}^{j}_{k/k-1} + K^{j}_{k}\left[Z_k - Z^{j}_{k/k-1}\right]$

Filter covariance: $p^{j}_{k/k} = \left[I - K^{j}_{k}H_j\right]p^{j}_{k/k-1}$

In the above:

$$H_j = \left(\frac{\partial h_1(\cdot)}{\partial \hat{x}^{j}_{k/k-1}}, \cdots, \frac{\partial h_N(\cdot)}{\partial \hat{x}^{j}_{k/k-1}}\right)^{T}$$

$$Z^{j}_{k/k-1} = \left[h_1\left(\hat{x}^{j}_{k/k-1}\right)^{T}, \cdots, h_N\left(\hat{x}^{j}_{k/k-1}\right)^{T}\right]^{T}$$

The third step is the calculation of the model update probability. For $j = 1, 2, \cdots r$, the corresponding model update probability expression is:

$$
\begin{aligned}
\mu^{j}_{k} &= P\{M_k = M_j/Z^k\} \\
&= \frac{1}{C}\Lambda^{j}_{k}\sum_{i=1}^{r}P\{M_k = M_j/M_{k-1} = M_i, Z^{k-1}\}P\{M_{k-1} = M_i/Z^{k-1}\} \\
&= \frac{1}{C}\Lambda^{j}_{k}\sum_{i=1}^{r}p_{ij}\mu^{i}_{k-1} \\
&= \frac{1}{C}\Lambda^{j}_{k}\overline{C}_{j}
\end{aligned}
\tag{4.37}
$$

where C is the normalization constant and $C = \sum_{i=1}^{r}\Lambda^{j}_{k}\overline{C}_{j}$. Set Λ^{j}_{k} as the likelihood function while observing Z_k. Its expression is:

$$
\begin{aligned}
\Lambda^{j}_{k} &= p\left(Z_k/M_k = M_j, Z^{k-1}\right) \\
&= \frac{1}{\sqrt{2\pi S^{j}_{k}}}\exp\left\{-\frac{1}{2}\left[Z_k - \hat{Z}^{j}_{k/k-1}\right]^{T}\left(S^{j}_{k}\right)^{-1}\left[Z_k - \hat{Z}^{j}_{k/k-1}\right]\right\}(\text{Assuming Gauss})
\end{aligned}
$$

where:

$$S^{j}_{k} = \text{diag}\left[\frac{\partial h_1(\cdot)}{\partial \hat{x}^{j}_{k/k-1}} \cdot p^{j}_{k/k-1} \cdot \frac{\partial h_1(\cdot)^{T}}{\partial \hat{x}^{j}_{k/k-1}} + R^{1}_{k/k-1}, \cdots,\right.$$

$$\left. + \frac{\partial h_N(\cdot)}{\partial \hat{x}^{j}_{k/k-1}} \cdot p^{j}_{k/k-1} \cdot \frac{\partial h_N(\cdot)^{T}}{\partial \hat{x}^{j}_{k/k-1}} + R^{N}_{k/k-1}\right].$$

The fourth step is the estimation fusion. The overall estimation and overall estimation error covariance matrix of the first Z_k moment is given by the following:

$$\hat{x}_{k/k} = \sum_{j=1}^{r} \hat{x}_{k/k}^{j} \mu_k^{j} \tag{4.38}$$

$$p_{k/k} = \sum_{j=1}^{r} \mu_k^{j} \left\{ p_{k/k}^{j} + \left[\hat{x}_{k/k}^{j} - \hat{x}_{k/k} \right] \left[\hat{x}_{k/k}^{j} - \hat{x}_{k/k} \right]^{T} \right\} \tag{4.39}$$

The above steps constitute a cycle of the IMM algorithm. By repeating the procedure, we can realize the adaptive tracking of maneuvering targets.

4.4 Tracking Filter Form

4.4.1 Collect Data

Under normal circumstances, the data can be obtained synchronously. The reason is for this is that, firstly, the system has enough communication ability. Secondly, a polynomial data smoothing device can be used for data pre-processing to time alignment. Set $z_{k+1,i}$ as the measurement of the first t_{k+1} moment of the $i(i = 1, 2, \cdots, m)$ the receiving station. Then:

$$z_{k+1,i} = h_i(X_{k+1}) + v_{k+1,i} \quad i = 1, 2, \cdots m \tag{4.40}$$

where $\{v_{k+1,i}\}$ is the white Gaussian noise sequence of the zero mean covariance matrix $R_{k+1,i}$, m are the given numbers of stations of the first t_{k+1} moment, and X_k is the target state vector.

All the measurement vectors constitute a new measurement vector:

$$z_{k+1} = \left[z_{k+1,1}^{T}, z_{k+1,2}^{T}, \cdots, z_{k+1,m}^{T} \right]^{T} \tag{4.41}$$

If $z_{k+1,i}$ is a column vector with n dimensions, then z_{k+1} is a column vector with $N = m \times n$ dimensions (otherwise, $M = n_1 + n_2 + \cdots + n_m$). If the measurement noise of each of the different stations is not associated with each other, then the covariance matrix of z_{k+1} is:

$$R_{k+1} = \begin{bmatrix} R_{k+1,1} & 0 & \cdot & \cdot & \cdot \\ 0 & R_{k+1,2} & & & \\ & & \cdot & & \\ & & & \cdot & \\ & & & & R_{k+1,m} \end{bmatrix} \tag{4.42}$$

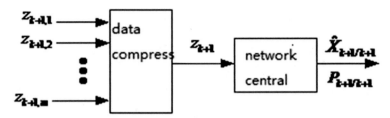

Fig. 4.1 Centralized processing structure

4.4.2   Centralized Processing Filter Form

The centralized processing structure is shown in Fig. 4.1. In the centralized processing structure, data provided and collected by the receiving stations can be parallel and serial processed, or similar data are compressed and then processed. In the linear case, three kinds of filter algorithms are consistent with the optimal filtering. When the measurement data statistics provided by receiving stations are independent, the computation efficiency of the data compression method is the highest, with serial processing taking second place, and parallel processing having a relatively poor performance.

4.4.2.1   Parallel Filtering

The parallel filtering method increases the dimension of the measurement vector of the filter, and then processes the filtering at the higher dimension, to, thus, estimate the states of the target. The method has no requirements for the measurement equation form of the receiving station. Even when the measurement errors of each terminal are related to each other, the method can also process them directly, so it is the most flexible to use. But because of the introduction of high dimension matrix multiplication and inversion, it brings with it a large amount of calculations. After applying formulas (4.41) and (4.42) for a series of operations in the filtering equation, we have the following.

   State:

$$\hat{X}_{k+1/k+1} = \hat{X}_{k+1/k} + \sum_{i=1}^{m} K_{k+1,i}\left(z_{k+1,i} - h_i\left(\hat{X}_{k+1/k}\right)\right) \tag{4.43}$$

Gain:

$$K_{k+1,i} = P_{k+1/k+1}H_{k+1,i}^T R_{k+1,i}^{-1} \tag{4.44}$$

Fig. 4.2 Parallel filtering

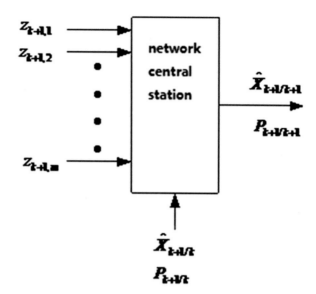

Covariance:

$$P_{k+1/k+1}^{-1} = P_{k+1/k}^{-1} + \sum_{i=1}^{m} \left(H_{k+1,i}^{T} R_{k+1,i}^{-1} H_{k+1,i} \right) \tag{4.45}$$

$H_{k+1,i}$ is the Jacobian matrix when $h_i(k+1)$ is at the state of $\hat{X}_{k+1/k}$

The parallel filtering structure of a network radar countermeasure system is shown in Fig. 4.2.

4.4.2.2   Serial Filter

In the serial filter method, one of the terminal's measurements is filtered regularly, and then the extrapolation time of the other terminals' measurements filtering is set to zero. Repeat updating the current target state.

State:

$$\hat{X}_{k+1/k+1,i} = \hat{X}_{k+1/k+1,i-1} + K_{k+1,i}\left(z_{k+1,i} - h_i\left(\hat{X}_{k+1/k+1,i-1} \right) \right) \tag{4.46}$$

Gain:

$$\begin{aligned} K_{k+l,i} &= P_{k+l/k+l,i-1} H_{k+l,i}^{T} \left(H_{k+l,i} P_{k+l/k+l,i-1} H_{k+l,i}^{T} + R_{k+l,i} \right)^{-1} \\ \text{or} \quad K_{k+l,i} &= P_{k+l/k+l,i} H_{k+l,i}^{T} R_{k+l,i}^{-1} \end{aligned} \tag{4.47}$$

Covariance:

$$P_{k+1/k+1,i} = P_{k+1/k+1,i-1} - K_{k+1,i} H_{k+1,i} P_{k+1/k+1,i-1} \tag{4.48}$$

$$\text{or} \quad P^{-1}_{k+1/k+1,i} = P^{-1}_{k+1/k+1,i-1} + H^T_{k+1,i} R^{-1}_{k+1,i} H_{k+1,i} \tag{4.49}$$

For $i = 1, 2, \cdots, m$:

$$\hat{X}_{k+1/k+1,0} = \hat{X}_{k+1/k}, \hat{X}_{k+1/k+1} = \hat{X}_{k+1/k+1,m} \tag{4.50}$$

$$P_{k+1/k+1,0} = P_{k+1/k}, P_{k+1/k+1} = P_{k+1/k+1,m} \tag{4.51}$$

The first i station of measurement is used in the first i time step to update the status. Calculate the state estimation after a first m time step:

$$\hat{X}_{k+1/k+1} = \hat{X}_{k+1/k} + \sum_{i=1}^{m} K_{k+1,i} \left(z_{k+1,i} - h_i \left(\hat{X}_{k+1/k+1,i-1} \right) \right) \tag{4.52}$$

For different m terminal measurement sets, this method takes m times of recursive filtering. In the process of each filter, the measurement matrix and measurement error covariance in the filter equation adaptively transform corresponding to the different receiving stations. The method has no limitations in regards to the measuring form of each station, but because the center station carries out filtering processing to each batch of station measurements, then when the measurement that the central station receives per unit time becomes larger, the consumption of filter computing resources will also be very large.

The serial filtering structure of a network radar countermeasure system is shown in Fig. 4.3.

4.4.2.3   Data Compression

The data compression filter method can realize the recombination of multi-station measurements according to a certain standard, and then filter the composite measure. The filtering method often lacks some flexibility, such as other additional conditions that require the same dimension of station measurement matrix.

Measurement after compression:

$$z_{k+1} = R_{k+1} \left(\sum_{i=1}^{m} R^{-1}_{k+1,i} z_{k+1,i} \right) \tag{4.53}$$

Covariance matrix:

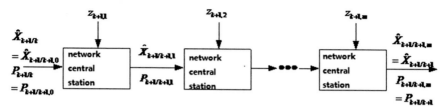

Fig. 4.3 Serial filtering

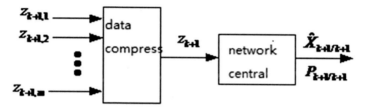

Fig. 4.4 Data compression filter

$$R_{k+1}^{-1} \sum_{i=1}^{m} R_{k+1,i}^{-1} \tag{4.54}$$

After obtaining formulas (4.53) and (4.54), updating the equation becomes the standard Kalman filtering equation.

The data compression filter structure of a network radar countermeasure system is shown in Fig. 4.4.

From the previous discussion, although the data compression method has high efficiency, it requires that the receiving station measurement matrices have the same dimensions. The measurement form of serial filtering to each station does not have any such limitations and its flexibility is strong. Meanwhile, because the network center has filter resources with strong data processing ability, it is not a difficult problem to solve. At the same time, the filtering efficiency of serial filtering is higher than that of parallel filtering, so serial filtering is mainly considered in centralized processing.

4.4.3 The Filter Form of Distributed Processing

The architecture of distributed processing is shown in Fig. 4.5.

Assume that, for the same target tracking, the measurement noises of different receiving stations are completely independent. At time k, the local estimation of each receiving station on target motion state X is $\hat{X}^i$. The corresponding error covariance matrix is P^{ii}, $i = 1, 2, \cdots, m$, and take P^{ij} as the cross-covariance matrix between the local estimation error of any two different receiving stations as i and j.

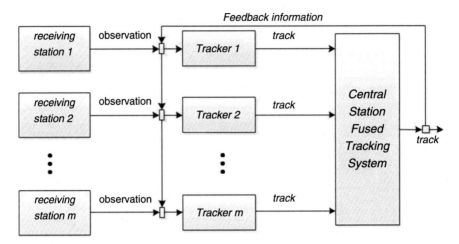

Fig. 4.5 Distributed structure of a network radar countermeasure system

We also assume that the system is a Gaussian system, and, in the sense of maximum likelihood (ML), the central station can construct a (negative) log-likelihood function as follows:

$$
L(X) = -\ln p\left(\hat{X}^1, \hat{X}^2, \cdots, \hat{X}^m \big| X\right)
$$

$$
= c + \frac{1}{2}\left\{ \begin{bmatrix} \hat{X}^1 \\ \hat{X}^2 \\ \vdots \\ \hat{X}^m \end{bmatrix} - \begin{bmatrix} I \\ I \\ \vdots \\ I \end{bmatrix} X \right\}^T P^{-1} \left\{ \begin{bmatrix} \hat{X}^1 \\ \hat{X}^2 \\ \vdots \\ \hat{X}^m \end{bmatrix} - \begin{bmatrix} I \\ I \\ \vdots \\ I \end{bmatrix} X \right\} \tag{4.55}
$$

where c is a constant, I is a unit matrix whose dimension is the same as that of the state vector X, and:

$$
P = \begin{bmatrix} P^{11} & P^{12} & \cdots & P^{1N} \\ P^{21} & P^{22} & \cdots & P^{2N} \\ \vdots & \vdots & \ddots & \vdots \\ P^{N1} & P^{N2} & \cdots & P^{NN} \end{bmatrix} \tag{4.56}
$$

When the central station completes the fusion without feedback information, then:

$$P_{k/k}^{ij} \overset{def}{=} E\left[\widetilde{X}_{k/k}^{i}\left(\widetilde{X}_{k/k}^{j}\right)^{T}\right]$$

$$= \left(I - K_k^i H_k^i\right)\left(F_{k-1}P_{k-1/k-1}^{ij}F_{k-1}^{T} + \Gamma_{k-1}Q_{k-1}\Gamma_{k-1}^{T}\right)\left(I - K_k^j H_k^j\right)^{T}$$

$$= P_{k/k}^{i}\left(P_{k/k-1}^{i}\right)^{-1}\left(F_{k-1}P_{k-1/k-1}^{ij}F_{k-1}^{T} + \Gamma_{k-1}Q_{k-1}\Gamma_{k-1}^{T}\right)\left(P_{k/k-1}^{j}\right)^{-1}P_{k/k}^{j}$$

$$(4.57)$$

where:

$$\widetilde{X}_{k/k}^{f} = X_k - \hat{X}_{k/k}^{f}$$

$$= F_{k-1}X_{k-1} + \Gamma_{k-1}W_{k-1} - F_{k-1}\hat{X}_{k-1/k-1}^{f} - K_k^{f}\left[H_k^{f}(F_{k-1}X_{k-1} + \Gamma_{k-1}W_{k-1})\right.$$

$$\left. + V_k^{f} - H_k^{f}F_{k-1}\hat{X}_{k-1/k-1}^{f}\right]$$

$$= \left(I - K_k^{f}H_k^{f}\right)F_{k-1}\widetilde{X}_{k-1/k-1}^{f} + \left(I - K_k^{f}H_k^{f}\right)\Gamma_{k-1}W_{k-1} - K_k^{f}V_k^{f}(f = i, j)$$

When the fusion is completed by the central station with feedback information, then:

$$P_{k/k}^{ij} = \left(I - K_k^i H_k^i\right)\left(F_{k-1}P_{k-1/k-1}F_{k-1}^{T} + \Gamma_{k-1}Q_{k-1}\Gamma_{k-1}^{T}\right)\left(I - K_k^j H_k^j\right)^{T}$$

$$= P_{k/k}^{i}P_{k/k-1}^{-1}\left(F_{k-1}P_{k-1/k-1}F_{k-1}^{T} + \Gamma_{k-1}Q_{k-1}\Gamma_{k-1}^{T}\right)P_{k/k-1}^{-1}P_{k/k}^{j} \qquad (4.58)$$

$$= P_{k/k}^{i}P_{k/k-1}^{-1}P_{k/k}^{j}$$

Both the fusion algorithm with feedback and the distributed fusion algorithm have the same performance. The main advantage of introducing feedback into the system is that the local estimation error covariance can be reduced.

Let $\xi \overset{def}{=} [I, I, \cdots, I]^{T}, \hat{\chi} \overset{def}{=} \left[\left(\hat{X}^1\right)^{T}, \left(\hat{X}^2\right)^{T}, \cdots, \left(\hat{X}^m\right)^{T}\right]^{T}$ and $\nabla_X L(X) = 0$, then we obtain the ML fusion estimation and the corresponding error covariance matrix of fusion estimation for the target's motion state X as follows:

$$\begin{cases} \hat{X}^{ML} = \left(\xi^T P^{-1}\xi\right)^{-1}\xi^T P^{-1}\hat{\chi} \\ P^{ML} = \left(\xi^T P^{-1}\xi\right)^{-1} \end{cases} \qquad (4.59)$$

If $P^{ij} = 0$(when $i \neq j$) in formula (4.57), i.e., when the estimation errors of all the receiving stations in the network radar countermeasure system are uncorrelated, then the fusion equation of formula (4.59) can be rewritten as follows:

$$\begin{cases} \hat{X} = \left[\displaystyle\sum_{i=1}^{m} (P^i)^{-1} \right]^{-1} \displaystyle\sum_{i=1}^{m} (P^i)^{-1} \hat{X}^i \\ P^{-1} = \displaystyle\sum_{i=1}^{m} (P^i)^{-1} \end{cases} \qquad (4.60)$$

4.5   Joint Probabilistic Data Association Algorithm

Multi-target tracking is an important function of the system, which plays an important role in many applications, such as ballistic missile defense, air defense (no-load early warning), air attack (multi-target attack), air traffic control (civil aircraft), aerial surveillance, radar tracking, etc.

The purpose of multi-target tracking is to decompose the measurement data that the detectors receive into observation sets or tracks produced by different corresponding target sources. Once the track is formed and confirmed, the number of targets tracked and their motion parameters, such as position, speed, acceleration, target classification features, etc. that correspond to each track can be estimated accordingly.

Elements involved in the issue of multiple targets include the formation of a tracking gate (associated area), data association and tracking maintenance, tracking initiation, and tracking termination. Among them, data association is the most important and difficult aspect of multi-target tracking technology. The basic concept of multi-target tracking was firstly presented by Wax in 1955. In 1964, Sittler made a breakthrough in multi-target tracking theory, including many aspects like data association. Until the early 1970s, multi-target tracking theory barely aroused the great interest of people. Works conducted by the Bar-Shalom and Singer have pushed modern multi-target tracking technology into further development, one sign of which is the combination of data correlation techniques and organic filter theory. In 1978, Bar-Shalom pointed out that multi-target tracking under instances of dense multi-echo was a field yet to be explored. But, so far, few research literatures on this topic have been produced, proving the difficulty of this issue. Naturally, multi-target tracking technology will undoubtedly become the focus of future study. The multi-target tracking of the network radar countermeasure systems can be divided into two parts: single station or network central station on multi-target tracking and multiple receiving stations on multi-target tracking.

In the network radar countermeasure system, due to the lack of prior knowledge based on tracking instances, and due to the limitations of receiving stations, the whole measurement process will inevitably introduce measurement errors (noise). In addition, due to the uncertainty in the exact number of targets, even if there is only one target, because of clutter jamming, there may also be a plurality of effective measurements, so it is often required to establish corresponding relationships between targets and measurements by statistical methods. It is more complex for a multi-target situation. At this point, it cannot be determined whether the measured data come from the targets in which we are interested or from false alarms or other objects. Especially when the tracks of several targets overlap, the solution to correct associations between targets and measurements has been facing serious challenges. It is just because of the existence of uncertainties and randomness in the observation process and multi-target tracking instances that the corresponding relationship between echo measurements and its target sources is undermined, so much so that it is necessary to use data association to come up with a better solution.

The data association process of network radar countermeasure systems is a process that determines the corresponding relationship between measured information received by stations and target sources. It is the most central and important aspect of multi-target tracking systems. Relevant papers have indicated that, no matter what kind of target tracking algorithm is used, they all take the correct result of data association as a precondition; that is to say, if the association goes wrong, then the estimation results will be wrong correspondingly.

Over the years, many scholars have deduced a large number of algorithms to solve the data association problem, and they can be summarized into the following two categories: (1) ML type data association filtering algorithms; (2) Bayesian type data filtering algorithms, wherein the likelihood ratio of observed sequences is the base of ML type data association filtering algorithms (depending on the target's cumulative time information/observed time series). The Bayesian rule-based data association filtering algorithm is mainly focused on the use of space cumulative information (spatial probabilistic data association), and the joint probabilistic data association algorithm belongs to the Bayesian type algorithms. As a sub-optimal Bayesian algorithm, the joint probabilistic data association algorithm only studies the latest confirming measured collection. The algorithm lists the permutations and combinations of all targets and measurements, and then selects a reasonable joint event according to a certain criteria and calculates the joint probability. The joint probabilistic data association algorithm takes the possibility of multiple measurements from other targets being in the same target-associated domain into account, solving the problem of having multi-target measurements in an associated domain. Meanwhile, the algorithm itself is more complicated and computationally intensive, and with the increase of the target number, the split of the confirming matrix could cause a combinatorial explosion. For these reasons, the implementation of the joint probabilistic data association algorithm in engineering is more difficult.

In order to make the algorithm easy to implement, many people have put forward several sub-optimal joint probabilistic data association algorithms; Robert Fitzerald proposed an empirical formula used to calculate the probability of association. Roecher proposed a sub-optimal Joint probabilistic data association algorithm whose sum of the weighted probabilities is equal to 1. Joint probabilistic data association algorithms are now recognized as one of the most ideal methods to be used for target tracking in cluttered instances.

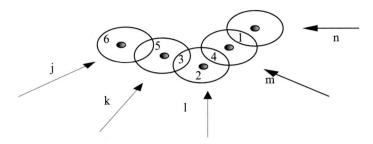

Fig. 4.6 Correspondence between the current track and new measurement

4.5.1 The Optimal Joint Probability Data Association Algorithm

The joint probabilistic data association algorithm is the core part of joint probability calculation. When calculating the joint probability, the possibility of measurements from other objects that fall into the tracking gate is considered; therefore, probability calculation is more complicated and computationally intensive. The calculation of joint probability is obtained by first calculating the probability of all possible joint events. One possible joint event at time k is expressed as $\theta_i(k)$. And one possible joint event is the non-contradiction association of the current track and measurement. Non-contradiction association is based on two basic assumptions:

1. Each measurement has a unique source;
2. For a given target, there is only up to one measurement that can take it as a source.

In addition, among those possible events, not every track has its corresponding measurement, which corresponds to a detection probability smaller than 1.

Figure 4.6 shows an example of a possible joint event. Assume that, currently, there are five tracks (j, k, l, m, n) and the new measurements obtained by the receiving station are $(1, 2, 3, 4, 5, 6)$. The numbers represent the measurement positions and the solid dots represent the positions of the tracking doors. One possible joint event is as follows:

$$\theta_i(k) = (k, 3)\ (l, 2)\ (j, 6)\ (n, 1)\ (\text{clutter point}\,4, 5) \tag{4.61}$$

This joint event indicates that tracks k, l, j and γ are, respectively, associated with the measurements 3, 2, 6, and 1. Track m was not detected in this scanning, and measurements 4 and 5 are clutters. In the joint probabilistic data association filters, we need to construct all possible joint events and calculate their probabilities. With the application of Bayes' rule, the conditional probability of a joint event based on all measurements Z^k that were taken until time k is:

$$P\{\theta_i(k)|Z^k\} = P\{\theta_i(k)|Z(k), Z^{k-1}\}$$
$$= \frac{1}{c}P[Z(k)|\theta_i(k), Z^{k-1}]P\{\theta_i(k)|Z^{k-1}\} \qquad (4.62)$$
$$= \frac{1}{c}P[Z(k)|\theta_i(k), Z^{k-1}]P\{\theta_i(k)\}$$

where the normalization constant $c = \sum_{i=1}^{n_k} P(Z(k)|\theta_i(k), Z^{k-1})P(\theta_i(k))$, with n_k the number of all joint events at time k and $z(k)$ is the measured collection at time k. $P\{\theta_i(k)|Z^{k-1}\}$ can be simplified to $P\{\theta_i(k)\}$. That is because of the assumption of possible joint events. The assumption is that the feasible joint probability of events at time k is unrelated to those at time $k - 1$ or all others before it. The first term on the right side of formula (4.62) is the current probability density function of the measurement collection. For a given current assumed joint event $\theta_i(k)$ and historical observation vector, the probability density function can be expressed as:

$$P[Z(k)|\theta_i(k), Z^{k-1}] = \prod_{j=1}^{m} P[z_j(k)|\theta_{j,i}(k), Z^{k-1}] \qquad (4.63)$$

where $z_j(k)$ is the jth measurement of $Z(k)$, then the term $\theta_{j,i}(k)$ indicates that the assignment of the jth measurement in this joint event comes from a current track or a clutter. m is the number of measurements that fall within the wave gate. The probability density function of each measurement can be expressed as:

$$P[z_j(k)|\theta_{j,i}(k), Z^{k-1}] = \begin{cases} N[v_j(k)] \text{ the value of } J \text{ is from target} \\ V^{-1} \text{ the value of } J \text{ is from clutter} \end{cases} \qquad (4.64)$$

where V is the volume of the extended wave gate, assuming that clutter obeys a uniform distribution. $N[]$ is a normal density function of the covariance matrix with zero mean value and its covariance is $v_j(k)$, where $v_j(k)$ is the difference between the jth measurement and the predicted position of the target at time k. Therefore:

$$P[Z(k)|\theta_i(k), Z^{k-1}] = V^{-\phi(\theta_i(k))} \prod_{j=1}^{m} [N[v_j(k)]]^{\tau_j(\theta_i(k))} \qquad (4.65)$$

In the joint events, if the jth measurement associates with a certain track, then $\tau_j(\theta_i(k))$ is equal to 1; otherwise, it is equal to 0. $\phi(\theta_i(k))$ is the amount of clutter in the joint event:

$$\phi(\theta_i(k)) = \sum_{j=1}^{m} (1 - \tau_j(\theta_i(k))) \qquad (4.66)$$

The next step will be to calculate the prior probability $P\{\theta_i(k)\}$. Analyses show that, once $\theta_i(k)$ is given, then the target detection indication $\delta(\theta_i(k))$ (which shows whether the target is detected within a feasible joint event) and the false measurement $\phi(\theta_i(k))$ can be completely determined. The expression of the prior probability can be written as:

$$
\begin{aligned}
P\{\theta_i(k)\} &= P\{\theta_i(k), \delta(\theta_i(k)), \phi(\theta_i(k))\} \\
&= P\{\theta_i(k) | \delta(\theta_i(k)), \phi(\theta_i(k))\} P\{\delta(\theta_i(k)), \phi(\theta_i(k))\}
\end{aligned}
\tag{4.67}
$$

In addition, it can also be noted that, when the false alarm measurement data are given, the joint event $\theta_i(k)$ will be uniquely determined by the target detection indication $\delta(\theta_i(k))$. The number of possible events that include $\phi(\theta_i(k))$ false alarms is $C_m^{\phi(\theta_i(k))}$. For the remaining $m - \phi(\theta_i(k))$ real measurements, among those events that contain $\phi(\theta_i(k))$ false alarms, there are $[m - \phi(\theta_i(k))]!$ possible target associations.

Therefore:

$$
\begin{aligned}
P\{\theta_i(k) | \delta(\theta_i(k)), \phi(\theta_i(k))\} &= \frac{1}{[m - \phi(\theta_i(k))]! C_m^{\phi(\theta_i(k))}} \\
&= \frac{\phi(\theta_i(k))!}{m!}
\end{aligned}
\tag{4.68}
$$

and:

$$
P\{\delta(\theta_i(k)), \phi(\theta_i(k))\} = \prod_{t=1}^{T} \left(P_D^t\right)^{\delta_t(\theta_i(k))} \left(1 - P_D^t\right)^{1 - \delta_t(\theta_i(k))} u_f(\phi(\theta_i(k)))
\tag{4.69}
$$

where P_D^t is the detection probability of track t. $\delta_t(\theta_i(k))$ is a binary indicator which indicates whether track t is associated with the measurement in a joint event $\theta_i(k)$. If track t is associated with the measurement, then $\delta_t(\theta_i(k))$ is equal to 1; otherwise, it is equal to 0. $u_f(\phi(\theta_i(k)))$ are the a priori probability allocation functions of the false alarm measurement data. According to the model used by the probability allocation function $u_f(\phi(\theta_i(k)))$, the parametric joint probability data association will use a Poisson distribution of $u_f(\phi(\theta_i(k)))$, and the non-parametric joint probability data association uses a uniform distribution of $u_f(\phi(\theta_i(k)))$; that is:

$$
u_f(\phi(\theta_i(k))) = \begin{cases} e^{-\lambda V} \dfrac{(\lambda V)^{\phi(\theta_i(k))}}{\phi(\theta_i(k))!}, & \textbf{parameter JPDA} \\ \varepsilon, & \textbf{nonparametric JPDA} \end{cases}
\tag{4.70}
$$

Accordingly, the prior probability of feasible event is:

$$P\{\theta_i(k)\} = \frac{\phi(\theta_i(k))!}{m!} u_f(\phi(\theta_i(k))) \prod_{t=1}^{T} \left(P_D^t\right)^{\delta_t(\theta_i(k))} \left(1 - P_D^t\right)^{1-\delta_t(\theta_i(k))} \quad (4.71)$$

When using the parametric model, it can be written as:

$$P\{\theta_i(k)|Z^k\} = \frac{\lambda^{\phi(\theta_i(k))}}{c} \prod_{t=1}^{m} \left[N[\nu_j(k)]\right]^{\tau_i(\theta_i(k))} \prod_{t=1}^{T} \left(P_D^t\right)^{\delta_t(\theta_i(k))} \left(1 - P_D^t\right)^{1-\delta_t(\theta_i(k))}$$

$$(4.72)$$

When using the non-parametric model, it can be written as:

$$P\{\theta_i(k)|Z^k\} = \frac{\phi(\theta_i(k))!}{c'V^{\phi(\theta_i(k))}} \prod_{j=1}^{m} \left[N[\nu_j(k)]\right]^{\tau_j(\theta_i(k))} \prod_{t=1}^{T} \left(P_D^t\right)^{\delta_t(\theta_i(k))} \left(1 - P_D^t\right)^{1-\delta_t(\theta_i(k))}$$

$$(4.73)$$

According to the total probability theorem, the association probability of track t and measurement j is equal to the sum of the association probability of track t and measurement j in all joint events. It can be expressed as:

$$\beta_j^t(k) = \sum_{\theta_i(k)} P\{\theta_i(k)|Z^k\} w(\theta_i(k)) \quad (4.74)$$

In the above expression, $w(\theta_i(k))$ is a binary variable which determines whether track t is associated with measurement j. If there is a certain association between them, then $w(\theta_i(k))$ is equal to 1; otherwise, it is 0.

After the association probability of the measurement and target is derived with the previous analysis, the main processes for the data association algorithm of the optimum joint probability are summarized as follows.

The initial value $\hat{X}^t(0/0)$, $P^t(0/0)$, $t = 1, 2, \cdots, T$

State prediction:

$$\hat{X}^t(k/k - 1) = F^t(k - 1)\hat{X}^t(k - 1/k - 1) \quad (4.75)$$

Echo prediction:

$$\hat{Z}^t(k/k - 1) = H^t(k)\hat{X}^t(k/k - 1) \quad (4.76)$$

Prediction of the covariance matrix:

$$\hat{P}^t(k/k - 1) = F^t(k - 1)P^t(k - 1/k - 1)[F^t(k - 1)]^T + Q^t(k - 1) \quad (4.77)$$

Innovation vector prediction:

$$V_j^t(k) = Z_j(k) - \hat{Z}^t(k/k - 1) \tag{4.78}$$

Tracking threshold (to obtain the echo):

$$g_t^2 t = 1, 2, \cdots, T \tag{4.79}$$

Generating feasible joint events θ_i, $i = 1, 2, \cdots, n_k$;
  Calculating the probability of feasible joint events using formula (4.62);
  Calculation of the association probability using formula (4.74);
  Kalman gain matrix:

$$\begin{aligned} K^t(k) &= P^t(k/k - 1)H^t(k)\left[H^t(k)P^t(k/k - 1)(H^t(k))^T + R^t(k)\right]^{-1} \\ &= P^t(k/k - 1)H^t(k)[S^t(k)]^{-1} \end{aligned} \tag{4.80}$$

Kalman filter equations:

$$\begin{aligned} \hat{X}_j^t(k/k) &= \hat{X}^t(k/k - 1) + K^t(k)V_j^t(k) \\ \hat{X}^t(k/k) &= \hat{X}^t(k/k - 1) + K^t(k)V^t(k) \end{aligned} \tag{4.81}$$

The covariance matrix of the filter:

$$\begin{aligned} P^t(k/k) = P^t(k/k - 1) &- \left(1 - \beta_0^t(k)\right)K^t(k)S^t(k)[K^t(k)]^T \\ &+ \sum_{j=0}^{m}\beta_j^t(k)\left[\hat{X}_j^t(k/k)\left(\hat{X}_j^t(k/k)\right)^T\right] - \hat{X}^t(k/k)\left(\hat{X}^t(k/k)\right)^T \end{aligned} \tag{4.82}$$

Let $k = k + 1$ and go to step (2):

$$\beta_0^t(k) = 1 - \sum_{j=1}^{m}\beta_j^t(k).$$

4.5.2 The Simple Joint Probabilistic Data Association Algorithm

The simple joint probabilistic data association filter was proposed by Fitzgerald. In this simple joint probabilistic data association filter, the associated probability of track t and measurement j is given by:

$$P_{tj} = \frac{G_{tj}}{S_t + S_j - G_{tj} + B} \tag{4.83}$$

where:

$$G_{tj} = N\left[v_j(k)\right]$$
$$S_t = \sum_{j=1}^{m} G_{tj}$$
$$S_j = \sum_{t=1}^{T} G_{tj}$$
$$B = \text{Constant. Depends on the density of clutter}$$

From (4.83), we can see that the empirical probability formula of the simple joint probabilistic data association has the characteristics of joint probabilistic data association calculation, i.e., measurements that only appear in a track-associated domain will have a heavy weight, and those overlapping in several track-associated domains will have a light weight. If several tracks extend along with measurement j, then the weight will increase due to the decreases of S_j. If a track has a number of measurements, then the weight will increase as S_t decreases. This simple algorithm gives a higher weight to measurements that are close to the predicted location and associate with a minimum number of tracks. In the presence of γ_i measurements and one target, formula (4.53) can give the correct estimation result $P_{tj} = 1/n$. In the case of completely unknown correspondence between the γ_i targets and γ_i measurements, the estimated probability that formula (4.53) gives is $P_{tj} = 1/(2n - 1)$, while the correct result is $P_{tj} = 1/n$.

Fitzgerald believes that, under normal circumstances (unless clutter is very large), $B = 0$ can give satisfactory results. In this case, B should be a constant, instead of being a variable for a function of measurement number or measurement density. Fitzgerald also believes that, when the target state updates, considering the processor load factor of the computer, only 2–3 measurements should be used with the highest probability. Furthermore, another reason for limiting the number of measurements is that the empirical probability may give a very high weight to a false measurement, so as to cause the uncontrolled growth of the covariance matrix in instances with high target density. It should also be noted that, for each track $\sum_{j=1}^{m} P_{tj} \neq 1$, if the sum of the weighted probabilities is not equal to 1, then it leads to a significant increase in the multi-target tracking error of the simple joint probabilistic data association algorithm, especially when the target tracks overlap.

Description of the processes for the simple joint probabilistic data association filter algorithm follows.

The initial value is $\hat{X}^t(0/0)$, $P^t(0/0)$, $t = 1, 2, \cdots, T$

State prediction:

$$\hat{X}^t(k/k - 1) = F^t(k - 1)\hat{X}^t(k - 1/k - 1) \tag{4.84}$$

Echo prediction:

$$\hat{Z}^t(k/k-1) = H^t(k)\hat{X}^t(k/k-1) \tag{4.85}$$

Prediction of the covariance matrix:

$$\hat{P}^t(k/k-1) = F^t(k-1)P^t(k-1/k-1)[F^t(k-1)]^T + Q^t(k-1) \tag{4.86}$$

Innovation vector prediction:

$$V_j^t(k) = Z_j(k) - \hat{Z}^t(k/k-1) \tag{4.87}$$

Tracking threshold (to obtain the echo):

$$g_t^2 t = 1, 2, \cdots, T \tag{4.88}$$

Calculation of the association probability using formula (4.83);
State estimation:

$$\hat{X}^t(k) = \hat{X}^t(k/k-1) + K^t(k)V^t(k) \tag{4.89}$$

where, $\hat{X}^t(k/k-1)$ is the predicted state vector;
$K^t(k)$ is the filter gain;
$V^t(k) = \sum_{j=1}^{m} P_{tj}(k)V_j^t(k)$ is the combined correction term.
Error covariance vector of state estimation:

$$P(k/k) = P_{t0}(k)\hat{P}^t(k/k-1) + [1 - P_{t0}(k)][I - K^t(k)H^t(k)]\hat{P}^t(k/k-1) + P(k) \tag{4.90}$$

where $\quad P(k) = K^t(k)\left[\sum_{j=1}^{m} P_{tj}(k)V_j^t(k)\left(V_j^t(k)\right)^T - V^t(k)(V^t(k))^T\right](K^t(k))^T, \quad$ and

$P_{t0}(k)$ is the probability of all measurements not being within the effective wave gate of track t.

Compared with the joint probability data association algorithm, the main advantage of the experience-based joint probability data association algorithm is reflected in the decreased computational complexity, while the disadvantage lies in loss of performance.

4.5.3   Associated Algorithm of the Joint Probability Data with the Probability-Weighted Summation Equal to 1

In the calculation of the associated probability, the simplified algorithm of the joint probability data can avoid the calculation of all possible joint events of interconnected density functions in the track and measure through the use of a single event, thereby greatly reducing the amount of computation required. A single event, shown by formula (4.83), substitutes the connection between track t and measure j. Related documentation suggested that the optimized algorithm with the probability-weighted summation equal to 1 uses part of the joint events instead of a single event case, and this is just like the simplified joint probabilistic data. In the algorithm part of the joint event is the connection between measure j_1 and track t_1 and measure j_2 and track t_2, and this assumes that these connections do not compete with each other. Tracks t_1 and t_2 have a common overlapped area, and they have a common measure, which is regarded as a joint event. The joint event is sub-optimal, because all joint events (as shown in formula (4.61) describe the best subset. The subset only takes two events into consideration, and replaces formulas (4.72) or (4.73) with $j = 1$ to $j = 2$. The specific operation steps are as follows:

1. For each track t, keep a record of all measurements falling within a list of interconnection capacities:

$$L_t = \text{The measuring table of contents interconnected with track t;} \qquad (4.91)$$

2. For each measurement j, keep a record of the list of tracks which are included in all interconnected areas:

$$L_j = \text{The measuring table of contents interconnected with measure j;} \qquad (4.92)$$

3. Calculate the probability of a sub-optimal event using the following steps:

   (a) Construct a table of contents of the tracks for all measures of each interconnected area of the tracks of all Internet domains to the set (but excluding the table of contents of track t):

   $$LOT = \left(U L_{j} j \in \text{interconneted area of the track} \notin t; \right) \qquad (4.93)$$

   (b) For all $t \in LOT$, solve:

   $$M_t = \max(G_{ht})h \in L_t h \neq j \qquad (4.94)$$

   If $M_t = 0$, it means that we can use measure j as long as there is only one measure in the interconnected area, so suppose:

   $$M_t = G_{jt} \qquad (4.95)$$

If $LOT = \phi$, there are no other interconnected areas which share the measure with the track, so suppose:

$$H_{jt} = G_{jt} \tag{4.96}$$

where H_{jt} is part of the joint case, otherwise suppose:

$$H_{jt} = G_{jt}\left(\sum_{t \in LOT} M_t\right) \tag{4.97}$$

4. A sub-optimal probability:

$$\beta_{jt} = \frac{H_{jt}}{B + \sum_j H_{jt}} \tag{4.98}$$

where B is a constant which depends on the density of the clutter. If it is not high, we can take $B = 0$

Important features of this equation:

1. In the case of one goal and n measures, the probability is correct, $LOT = \phi$;
2. In the case of two goals and two measurements, the probability is also correct, and the probability is equal to 1/2;
3. For the cases when n goals and n measures completely overlapped, the probability is also correct and is equal to $1/n$;
4. Formula (4.83) considers that the measurement included in this track will be expanded by the other tracks, and formula (4.99) considers that tracks which share the measure with this track will also be extended by the further measurements.
5. For each track, the sum of the probabilities associated with the measure and the probability undetected by the track is equal to 1:

$$\sum_j P_{tj} = \frac{B}{B + \sum_j H_{tj}} + \sum_j \frac{H_{tj}}{B + \sum_j H_{tj}} \tag{4.99}$$
$$= 1$$

After obtaining the associated probabilities, we can filter the calculation with reference to formulas (4.75), (4.76), (4.77), (4.78), (4.79), (4.80), (4.81), and (4.82).

4.5.4   Improved Associated Algorithm of the Joint Probability Data

From the above analysis of the algorithm, it can be seen that due to taking into account part of possible joint events, capability of the associated algorithm of the joint probability data with the probability-weighted incidents summation equal to 1 is better than the simplified joint probability data related to a single incident, but worse than the optimal joint probability data. Formula (4.94) shows that, since the formula only considers the measure of the likelihood function with the most effective goal connection, in the situations where there is a large amount of clutter and multiple targets, it is easy to produce errors, and it leads to mistakes in formulas (4.98) and (4.99). For these cases, using the algorithm where the sum of the probabilities associated with the measure and the probability undetected by the track equals 1, the following improvements can be made: When $LOT \neq \phi$, if excluding measure j, the quantity of other measures falling outside the goal is larger than or equal to 4:

$$M_t = \text{sum}(G_{h_i t})/4 \, h_i \in L_t \, h_i \neq j \, i = 1, 2, 3, 4 \tag{4.100}$$

where $G_{h_i t} = i = 1, 2, 3, 4$ is the likelihood function value of the largest four measures. Otherwise:

$$M_t = \text{sum}(G_{h_i t})/n \, h_i \in L_t \, h_i \neq j \, i = 1, 2, \cdots, n(n < 4) \tag{4.101}$$

where n are the measures falling outside the goal, excluding measure j.

Equations (4.100) and (4.101): when only one measure excluding measure j falls at the determined door, the result is the same as that of formula (4.95). When many measures excluding measure j fall at the determined door, in order to reduce errors of the associated probability, take the summation of the effective likelihood functions of all the measures excluding measure j; this does not affect the correction of the result. Compared with the original algorithm, the improved algorithm differs at characteristic 4. The original algorithm considers that other tracks which share the measure with this track are also expanded by the only measure, but the improved algorithm considers that the other tracks which share the measure with this track also fall at the determined door and are expanded by other measures, excluding measure j. It means that the improved algorithm considers more events excluding the section joint events, so the capacity is between the optimal associated algorithm and the algorithm probability-weighted summation equal to 1, which is better than the simplified associated algorithm. From the standpoint of the amount of computation, the improved algorithm just adds a few more laws; therefore, it does not affect the capacity. Comparisons of examples are presented. Consider three tracks and four measures, as shown in Fig. 4.7.

Fig. 4.7 The relationship
among measures and
corresponding tracks

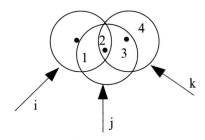

(i, j, k) are the three tracks, $(1, 2, 3, 4)$ are the four measures, and the points represent the locations of the tracking gates. The relationship among measures and corresponding tracks is:

$$\begin{aligned} i &- 1, 2 \\ j &- 1, 2, 3 \\ k &- 2, 3, 4 \end{aligned} \tag{4.102}$$

The optimal joint hypothesis is assumed when there exists eight tracks and their corresponding measures are:

$$\begin{aligned} &H1 : (i1), (j2), (k3) \quad H2 : (i1), (j2), (k4) \\ &H3 : (i1), (j3), (k2) \quad H4 : (i1), (j3), (k4) \\ &H5 : (i2), (j1), (k3) \quad H6 : (i2), (j1), (k4) \\ &H7 : (i2), (j3), (k1) \quad H8 : (i2), (j3), (k4) \end{aligned} \tag{4.103}$$

For the optimal parameters of the equation, when the detection probability is close to 1 (here, clutter is negligible), the optimized probability track i and measure 1 are:

$$P_{i1} = \frac{G_{i1}G_{j2}G_{k3} + G_{i1}G_{j2}G_{k4} + G_{i1}G_{j3}G_{k2} + G_{i1}G_{j3}G_{k4}}{G_{i1}G_{j2}G_{k3} + G_{i1}G_{j2}G_{k4} + G_{i1}G_{j3}G_{k2} + G_{i1}G_{j3}G_{k4} + G_{i2}G_{j1}G_{k3} + G_{i2}G_{j1}G_{k4}}$$
$$+ G_{i2}G_{j3}G_{k1} + G_{i2}G_{j3}G_{k4} \tag{4.104}$$

The probability of the simplified algorithm can be obtained from formula (4.83):

$$P_{i1} = \frac{G_{i1}}{G_{i1} + G_{i2} + G_{j1}} \tag{4.105}$$

where $B = 0$.

The probability of the associated algorithm probability-weighted summation equal to 1 can be obtained from formulas (4.91), (4.92), (4.93), (4.94), (4.95), (4.96), (4.97), (4.98), and (4.99):

$$P_{i1} = \frac{G_{i1}G_{j2} + G_{i1}G_{k3}}{G_{i1}G_{j2} + G_{i1}G_{k3} + G_{i2}G_{j1} + G_{i2}G_{k3}} \tag{4.106}$$

The probability of the improved associated algorithm can be obtained from formula (4.101):

$$P_{i1} = \frac{G_{i1}G_{j2} + G_{i1}G_{k3}}{G_{i1}G_{j2} + G_{i1}\left(\frac{G_{k2}+G_{k3}+G_{k4}}{3}\right) + G_{i2}G_{j1} + G_{i2}\left(\frac{G_{k3}+G_{k4}}{2}\right)} \tag{4.107}$$

4.6 Tracking Multiple Targets by Multiple Receiving Stations

During the process of tracking multiple targets by multiple receiving stations in the system, using a number of receiving stations to track multiple targets is the issue to be solved. Due to the differences in the multiple receiving stations during their work and in their properties, the observation numbers or data types are different as well. Or the data types can be the same but the observation data errors are different. There are two different system structures in a multi-target tracking system, namely, centralized and distributed. The corresponding algorithm ideas are: multi-station data with the algorithm and sequence of the serial multi-station data with the algorithm.

4.6.1 The Joint Probabilistic Data Association Algorithm of Parallel Multiple Receiving Stations

The joint probabilistic data association algorithm of parallel multiple receiving stations is an operational algorithm used in the centralized tracking system of network radar systems.

Assuming there are N_s receiving stations in a network radar countermeasure system, and the measurement errors among receiving stations are statistically independent. We first confirm that the measurement number from receiving station i in k moment is $m_{i,k}(i = 1, 2, \cdots N_s)$. For $1 \le t \le T$ and $L = (l_1, l_2, \cdots, l_{N_s})$ in which $0 \le l_1 \le m_{1,k}, \cdots, 0 \le l_{N_s} \le m_{N_s,k}$, suppose $\theta_L^t(k)$ is the real measurement number from receiving station i. Suppose $\beta_L^t(k)$ is expressed in a given measurement group of nations, circumstances, events, etc. with conditional probability:

$$\beta_L^t(k) = \prod_{i=1}^{N_s} \beta_{l_i,i}^t(k) \tag{4.108}$$

$\beta_{l_i,i}^t(k)$ is the single receiving station association probability deduced in Sect. 4.5.

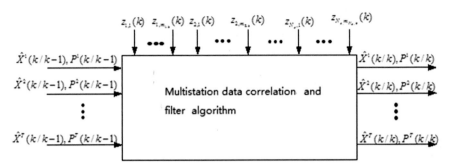

Fig. 4.8 The algorithm structure of joint probability data association of parallel multiple receiving stations

Therefore, based on multiple receiving stations, the target state for t is estimated to be:

$$\hat{X}^t(k/k) = \sum_L \beta_L^t(k)\hat{X}_L^t(k/k) \tag{4.109}$$

In this equation, $\hat{X}_L^t(k/k)$ expresses the target state estimation on the basis of the given measurement configuration, and the calculation formula is showed as below:

$$\hat{X}_L^t(k/k) = \hat{X}^t(k/k-1) + \sum_{i=1}^{N_s} K_i^t(k)\left[z_{l_i}^i(k) - H_i(k)\hat{X}^t(k/k-1)\right] \tag{4.110}$$

In the above equation, $\hat{X}^t(k/k-1)$ is the predicted state and $K_i^t(k)$ is the gain matrix obtained by filtering target t by receiving station i. The corresponding new covariance matrix is:

$$P^t(k/k) = \sum_L \beta_L^t(k)\left[P_L^t(k/k) + \hat{X}_L^t(k/k)\hat{X}_L^t(k/k)^T\right] \\ - \hat{X}^t(k/k)\hat{X}^t(k/k)^T \tag{4.111}$$

In this equation, $P_L^t(k/k)$ is the covariance according to the predicted state $\hat{X}_L^t(k/k)$. The structure chart is shown in Fig. 4.8

4.6.2   Joint Probabilistic Data Association Algorithm of Ordered Multiple Receiving Stations

The joint probabilistic data association algorithm of ordered multiple receiving stations is used in the network radar countermeasure distributed tracking system. In this algorithm, after using data related to the measurement information, the measure of the first receiving station is used to calculate the intermediate state estimation $\hat{X}_1^t(k/k)$ and corresponding covariance of every target. Then, we use the data association again and improve the intermediate state estimation by using the measure of the next receiving station. This is repeated until processing has been completed for the measure of the N_s receiving station. With this method, process the measures from each receiving station in turn. The algorithm structure is shown as below:

Take two receiving stations for example. The sequence of the serial processing of the multiple receiving stations' data on the filtering algorithm is as follows:

First step: For target t, use the state estimates $\hat{X}^t(k-1/k-1)$ and covariance matrix $P^t(k-1/k-1)$ of time $k-1$, then calculate the state prediction $\hat{X}^t(k/k-1)$, its $P^t(k/k-1)$, and the measuring forecast $\hat{Z}_1^t(k/k-1)$ of the first receiving station and corresponding covariance $S_1^t(k)$.

Second step: Manage the data correlation of the effective measure and the known track of the first receiving station.

Third step: Process the state estimation by using the confirmed measurement from step two to get the intermediate state estimation $\hat{X}_1^t(k/k)$ and its covariance $P_1^t(k/k)$.

Fourth step: Process the state estimation by using the confirmed measurement from step two and replace $\hat{X}_1^t(k/k-1)$ and its $P_1^t(k/k-1)$ by $\hat{X}_1^t(k/k)$ and $P_1^t(k/k)$. Calculate the measuring forecast $\hat{Z}_2^t(k/k-1)$ and corresponding covariance $S_2^t(k)$ of the second receiving station.

Fifth step: Manage the data correlation of the effective measure and the known track of the second receiving station.

Sixth step: Process the state estimation by using the confirmed measurement from step two to get the intermediate state estimation $\hat{X}^t(k/k)$ and its covariance $P^t(k/k)$.

Based on multi-target tracking data from the algorithm, as the receiving station number and the clutter density increases, the calculation time of the previously mentioned two types of algorithms will suddenly increase. Current researches indicate that the ordered multiple stations data association algorithm is preferable to the parallel multiple stations data association algorithm. This is because, in the parallel algorithm, the association and filtering of all measurements from the receiving stations at the current moment are gained by the state and covariance, which were estimated shortly before. This is compared to the ordered multiple stations data association algorithm, which uses the measurement information of the first receiving station to gain the association and filtering. It is good for gaining a better state estimation to process the association and filtering of the next receiving station. Hence, the use of the ordered multiple stations data association algorithm is more common. Now, process a simulation following the algorithm structure shown in Fig. 4.9.

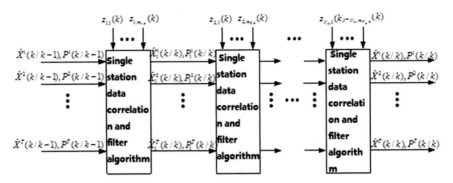

Fig. 4.9 Series of multi-station joint probability data relating to the algorithm structure

4.6.3 Simulation and Analysis

4.6.3.1 Simulation Example One

Let launching station coordinate be $T(-10, 0, 0.05)$ km, and receiving station coordinate be $R(30, 0, 0)$ km. Measurements of receiving station are distance, azimuth and elevation. Three targets move in uniform linear motion. Initial position of target 1 is $(50000, 50000, 10000)$ m, initial position of target 2 is $(51200, 50000, 10000)$ m, and initial position of target 3 is $(52400, 50000, 10000)$ m. All of the targets move in the speed of 100 m/s. The angles between the initial directions and X axis are $80°$, $90°$ and $100°$; Sampling period $T = 4$ s; The simulation time is 120 s; Both of standard deviations of the target location error at initialization state in the x and y direction are 80 m, which in the z direction is 10 m. The standard deviation of distance measurement error is 3 m, the standard deviation of azimuth and pitching angle measuring error is 3 mrad; and site error is 1 m. To compare the performance of the algorithm, we use uniformly accelerated motion model as track model. Target's state vector $X = [x, \dot{x}, y, \dot{y}, z, \dot{z}, \ddot{x}, \ddot{y}, \ddot{z}]^T$; The three targets' initial state vectors are:

$$X_1 = [50087, 19, 50374, 98, 10035, 0, 2, 2, 2]^T,$$
$$X_2 = [51176, 0, 50199, 102, 9994, 0, 2, 2, 2]^T,$$
$$X_3 = [52336, -19, 50462, 100, 10014, 0, 2, 2, 2]^T$$

The initial state error covariance matrix is:

$$P = \begin{bmatrix} 6400 & 1600 & 0 & 0 & 0 & 0 & 400 & 0 & 0 \\ 1600 & 800 & 0 & 0 & 0 & 0 & 300 & 0 & 0 \\ 0 & 0 & 6400 & 1600 & 0 & 0 & 0 & 400 & 0 \\ 0 & 0 & 1600 & 800 & 0 & 0 & 0 & 300 & 0 \\ 0 & 0 & 0 & 0 & 100 & 25 & 0 & 0 & 6.25 \\ 0 & 0 & 0 & 0 & 25 & 12.5 & 0 & 0 & 4.6875 \\ 400 & 300 & 0 & 0 & 0 & 0 & 150 & 0 & 0 \\ 0 & 0 & 400 & 300 & 0 & 0 & 0 & 150 & 0 \\ 0 & 0 & 0 & 0 & 6.25 & 4.6875 & 0 & 0 & 2.3438 \end{bmatrix};$$

We use simple joint probability data association algorithm, suboptimal joint probability data association algorithm whose weighted sum of probability is equal to 1 and improved algorithm to complete the tracking of these three cross movement targets. Monte Carlo simulation is completed for 10 times.

Set tracking parameter $\gamma = 16$, with measuring dimension $n_z = 3$, and check the χ^2 probability distribution table to find that the available door is $P_G = 0.9989$. The volume of the ellipsoidal tracking gate is:

$$V = 4\pi/3 \sqrt{|S(k)|} \gamma^{3/2} \qquad (4.112)$$

where $S(k)$ is the new rate of covariance. False measurement is in the right measurement as the center of a uniform cube with volume $A = n_c/\lambda \approx 10V$, where λ is a false alarm space density (per unit volume on the number of false measurements), $\lambda = 0.9$, n_c is the false measuring total, namely, $n_c = INT[10V\lambda + 1]$, which take the largest integer $INT[x]$ no greater than x. Expressed in RND as a set of uniformly distributed random numbers, the first x represents a false position of measurement area as follows:

$$x_i = a + (b - a)RND, \quad i = 1, 2, \cdots, n_c$$
$$y_i = c + (d - c)RND, \quad i = 1, 2, \cdots, n_c$$
$$z_i = e + (f - e)RND, \quad i = 1, 2, \cdots, n_c$$
$$a = x_k - q, b = x_k + q, c = y_k - q, d = y_k + q, e = z_k - q, f = z_k + q,$$

(x_k, y_k, z_k) is the correct measurement of position and $q = \sqrt[3]{10V}/2$. A volume of $A \approx 10V$ produces a large number of false measurements. By confirming the number of measurements within the region V, λV approximately obeys a Poisson distribution. The simulation results are shown in the following figures:

The operation times for the three kinds of algorithm are shown in Table 4.1.

The simulation result shows that: when we use simple JPDA algorithm, the maximum number of false measurements in target one's tracking gate is 21, this number in target two's tracking gate is 57, and this number in target three's tracking gate is 20. When we use the association algorithm whose sum of weighted probability is equal to 1, the maximum number of false measurements in target one's tracking gate is 21, this number in target two's tracking gate is 57, and this number

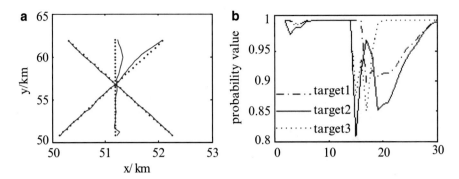

Fig. 4.10 (**a**) Simple joint probability correlation algorithm tracking results. (**b**) The corresponding correlation probability

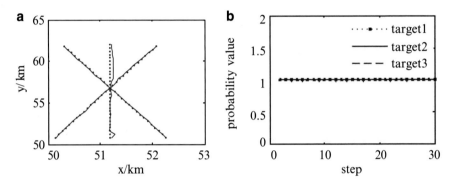

Fig. 4.11 (**a**) Results when the tracking weights sum to 1. (**b**) The corresponding correlation algorithm probability

in target three's tracking gate is 20. When we use the improved algorithm, the maximum number of false measurements in target one's tracking gate is 21, this number in target two's tracking gate is 56, and this number in target three's tracking gate is 20.

In Figs. 4.10, 4.11 and 4.12, we use midpoint line to simulate the target's real track, and the solid line to simulate the filtering track.

Figures 4.11 and 4.12 show a network radar countermeasure system based on distance and azimuth. The angle and elevation angle measurements combined with the probability weights summing to 1 data correlation algorithm and improved algorithm can accurately track the moving targets.

In Figs. 4.10, 4.11, and 4.12, we use simple joint probability data association algorithm and association algorithm whose probability weighted sum is equal to 1 to track the multi targets. In simple JPDA algorithm, position RMS error value where weights sum isn't equal to 1 is far greater than this value where weights sum

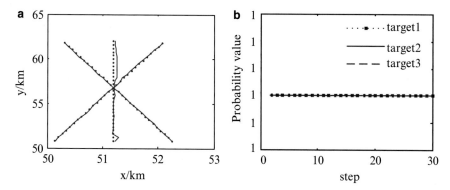

Fig. 4.12 (**a**) The improved algorithm tracking results. (**b**) The corresponding correlation probability

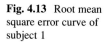

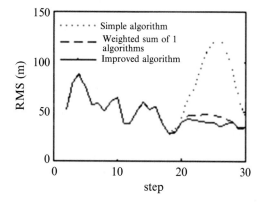

Fig. 4.13 Root mean square error curve of subject 1

is equal to 1. The reason is that in this condition, the information is not enough in the joint probability data association algorithm, which leads to the error of the target tracking, and reflects on the tracking of the two targets of the initial motion direction and the X direction of 80° and 90°.

As shown in Fig. 4.12, the improved algorithm inherits the original probability weighting algorithm and the characteristic of the weights being equal to 1. At the same time, as shown in Figs. 4.10, 4.11, 4.12, 4.13, 4.14, and 4.15, in the presence of a lot of clutter, the errors of the improved algorithm to obtain the three target positions have smaller RMS values, especially for goals 1 and 2. The reason for this is because the probabilities are equal to 1 and the correlation algorithm is considered part of the joint events on the basis of further increases in the number of events, so the tracking performance is closer to the optimal joint probability data association algorithm of the tracking performance.

Fig. 4.14 Root mean square error curve of subject 2

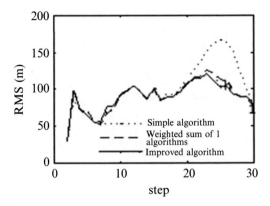

Fig. 4.15 Root mean square error curve of subject 3

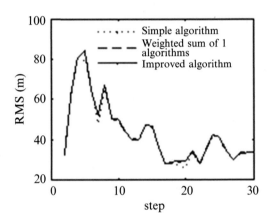

It can be seen from Table 4.1 that the computational complexities were similar for the three kinds of algorithm, especially the improved algorithm with probability weighted algorithm and the sum of the weights being equal to 1.

4.6.3.2 Simulation Example Two

The coordinates of launching station 1 are $T_1(-10, 0, 0.05)$km, the coordinates of launching station 2 are $T_2(10, 0, 0.05)$km, the coordinates of launching station 3 are $T_3(0, -10, 0.05)$km, and the coordinates of the receiving station are $R(30, 0, 0)$km, which measures the range sum of three stations. The initial position error in direction x, y, z is 60 m, 60 m, 10 m. The initial values of the three goals are:

$$X_1 = [50087, 19, 50374, 10035, 0, 2, 2, 2]^T,$$
$$X_2 = [51176, 0, 50199, 102, 9994, 0, 2, 2, 2]^T,$$
$$X_3 = [52336, -18, 50462, 100, 10014, 0, 2, 2, 2]^T,$$

The initial state error covariance matrix is:

Table 4.1 The operation times of the three kinds of algorithm

Algorithm	Simple joint probability data association algorithm	The algorithm when the probability weight sum is equal to 1	Improved algorithm
Operation time (s)	2.224	2.655	2.777
	2.231	2.735	2.753
	2.077	2.689	2.734
	2.316	2.638	2.546
	1.954	2.482	2.502
	1.94	2.425	2.463
	1.981	2.389	2.49
	2.127	2.497	2.578
	2.174	2.518	2.449
	2.112	2.509	2.613
Average operation time (s)	2.114	2.554	2.591

$$P = \begin{bmatrix} 3600 & 900 & 0 & 0 & 0 & 0 & 225 & 0 & 0 \\ 900 & 450 & 0 & 0 & 0 & 0 & 168.75 & 0 & 0 \\ 0 & 0 & 3600 & 900 & 0 & 0 & 0 & 225 & 0 \\ 0 & 0 & 900 & 450 & 0 & 0 & 0 & 168.75 & 0 \\ 0 & 0 & 0 & 0 & 100 & 25 & 0 & 0 & 6.25 \\ 0 & 0 & 0 & 0 & 25 & 12.5 & 0 & 0 & 4.6875 \\ 225 & 168.75 & 0 & 0 & 0 & 0 & 84.375 & 0 & 0 \\ 0 & 0 & 225 & 168.75 & 0 & 0 & 0 & 84.375 & 0 \\ 0 & 0 & 0 & 0 & 6.25 & 4.6875 & 0 & 0 & 2.3438 \end{bmatrix} ;$$

Noise is produced during the processing of parameter $\lambda = 10^{-6}$. The rest of the simulation conditions are the same as in simulation example one. The simulation results are shown in the following figures:

By the simulation calculation, the simple joint probabilistic data association algorithm in the tracking gate of target 1 reaches a maximum number of false measurements of 267, the largest false measurements number in the tracking gate of target 2 is 243, and the tracking gate in target 3 produced the largest number of false measurements of 166. Using the probability-weighted and correlation algorithm equal to 1 in the target 1 tracking door when the maximum number of false measurements was 171, the largest false measurements number in the tracking gate of target 2 was 169, and the tracking gate of target 3 produced the largest number of false measurements of 166. With the improved algorithm, the largest false measurements number in the tracking gate of target 1 was 171, the largest false measurements number in the tracking gate of target 2 was 169, and the largest number of false measurements in the tracking gate of target 3 was 166.

In Figs. 4.16, 4.17, 4.18, 4.19, 4.20, and 4.21, the dotted lines show the real trajectories, and the solid lines show the filtering tracks.

As shown in Figs. 4.16, 4.17, and 4.18, the network radar countermeasure system and measure of a single receiving station based on the three distances using the

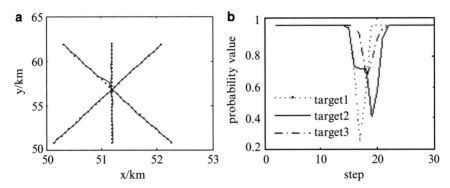

Fig. 4.16 (**a**) Simple joint probability correlation tracking results. (**b**) The corresponding algorithm correlation probability

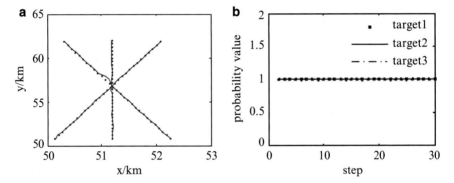

Fig. 4.17 (**a**) Results when the tracking correlation algorithm weight sum is equal to 1. (**b**) The corresponding correlation probability

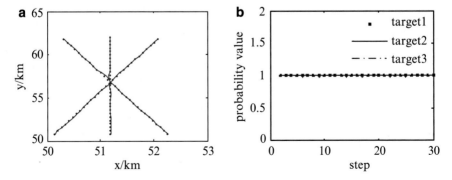

Fig. 4.18 (**a**) The improved algorithm tracking results. (**b**) The corresponding correlation probability

Fig. 4.19 Root mean square error curve of subject 1

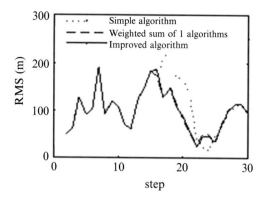

Fig. 4.20 Root mean square error curve of subject 2

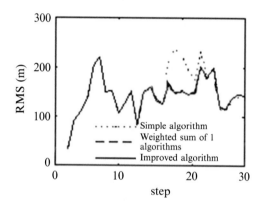

Fig. 4.21 Root mean square error curve of subject 3

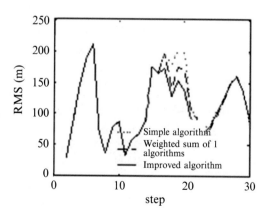

simple joint probability data association algorithm and probability weighting and equal to 1 data association algorithm and improved algorithm can better finish the cross-movement target tracking. Improved tracking performance of the algorithm is better than the weighted and tracking performance of the algorithm is equal to

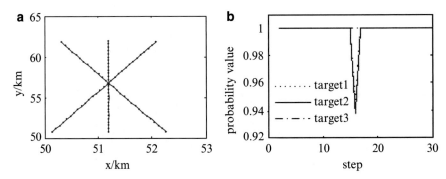

Fig. 4.22 (**a**) Simple joint probability correlation algorithm tracking results. (**b**) The corresponding correlation probability

1, mainly reflected in the target 3 tracking, and both are significantly better than simple tracking performance of joint probability data association algorithm.

4.6.3.3 Simulation Example Three

Passive mode time: network $R_0 = (0, 0, 0)$km, the coordinates of center station 1 are $R_1 = (30, 0, 0)$km, the coordinates of terminal 2 are $R_2 = (0, -30, 0)$km, and the coordinates of terminal 3 are $R_3 = (-21.2, 21.2, 0)$km. The standard deviation of the measurement errors for time is 3 ns. The electromagnetic wave propagation speed is $c = 3 \times 10^8$ m/s. Noise is produced during the processing of parameter $\lambda = 4 \times 10^{19}$. The rest of the simulation conditions are the same as in simulation example two. The simulation results are shown in the following figures:

By the simulation calculation, the simple joint probability data association algorithm in the tracking gate of target 1 reaches a maximum number of false measurements of 28, one of the largest false measurements number in the tracking gate of target 2 is 27, and one of the largest false measurements number in the tracking gate of goal 3 is 25. Using the probability-weighted and correlation algorithm equal to 1 in the target 1 tracking door when the maximum number of false measurements was 28, one of the largest false measurement numbers in the tracking gate of target 2 was 27, and one of the largest false measurement numbers in the tracking gate of goal 3 was 25.

The dotted lines in Figs. 4.22, 4.23, 4.24, 4.25, and 4.26 represent the target real trajectories, and the solid lines show the filtering tracks.

As shown in Figs. 4.22 and 4.23, based on time difference measuring using probability-weighted and equal to 1 correlation algorithm and simple joint probability data association algorithm can better finish the cross-movement target tracking. Based on time difference measuring using probability-weighted and equal to 1 of the correlation algorithm is easy joint probability of cross-correlation algorithm of three target tracking, in most of the time period for the position of smaller RMS

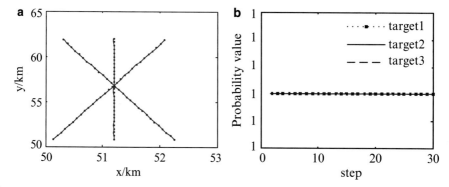

Fig. 4.23 (**a**) Results of tracking correlation algorithm weight sum is equal to 1. (**b**) The corresponding correlation probability

Fig. 4.24 Root mean square error curve of subject 1

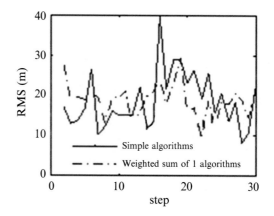

Fig. 4.25 Root mean square error curve of subject 2

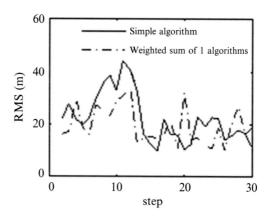

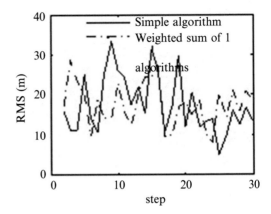

Fig. 4.26 Root mean square error curve of subject 3

error values. But from the perspective of the peak and the convergence value of location RMS error, both can satisfy the requirement of track correlation algorithm.

Conclusion For a network radar countermeasure system with a single receiving station in active work mode and the center working in passive mode, the cases of more intense echo cross to track a moving target, using the algorithm of probability weighted and equal to 1 simple tracking effect is much better than that of the joint probability data association algorithm. The improved algorithm based on probability weighted and equal to 1 is due to consider more joint events, thus obtained than probability weighted and correlation algorithm and better tracking performance is equal to 1, and from the perspective of the computational complexity of algorithm, probability weighted sum is equal to 1 algorithm and improved algorithm of simple computation and joint probability data association algorithm computation were similar, meet the needs of engineering practice.

4.6.3.4   Simulation Example Four

The transmitter station has coordinates $T(-10, 0, 0.05)$km and the receiver station has coordinates $R(30, 0, 0)$km. The measuring amount of the receiving station is the distance sum, orientation, and pitching. The initial positions of targets 1, 2, and 3 are $(50000, 50000, 10000)$m, $(51200, 50000, 10000)$m, and $(54300, 50000, 10000)$m, respectively.

The movement velocity of target 1 is 80 m/s, that of target 2 is 80 m/s, and that of target 3 is 90 m/s. Targets 1 and 2 move in uniform linear motion and the initial motion directions are both at 70° angles to the X axis, with target 3 moving in a linear motion with initial motion at 0–40 s, and the initial motion direction making an angle of 120° with the X axis. It makes a circular motion at 40–80 s, with movement radius 20 m, centripetal acceleration 0.003 m/s², and moves in a linear motion with initial motion at 80–120 s.

The sampling period T = 4 s, the simulation time is 120 s, and the initial positioning errors of x, y, z are 50 m, 50 m, 10 m. The standard deviations of distance and measurement error are 3 m, the standard deviations of azimuth and pitch angle measurement errors are both 3 mrad, and the site error is 1 m. The tracking model of targets 1 and 2 use uniformly accelerated motion, while target 3 uses the Singer model. The parameter settings of the Singer model are: maneuvering frequency 1/20, maxima maneuvering acceleration 5 m/s^2, the probability of maneuvering equal to the maximum value of acceleration is 0.95, probability of equal to zero 0.05. The state vector of the target is $X = [x, \dot{x}, y, \dot{y}, z, \dot{z}, \ddot{x}, \ddot{y}, \ddot{z}]^T$, and the initial values of the three targets are:

$$X_1 = [50134, 28, 50375, 75, 10022, 0, 2, 2, 2]^T,$$

$$X_2 = [51179, 27, 50379, 75, 9951, 0, 2, 2, 2]^T,$$

$$X_3 = [54127, -45, 50300, 78, 9967, 0, 2, 2, 2]^T;$$

The simulation results are as follows:

From the simulation, while using the simple joint probability data association algorithm, the maximum amounts of targets 1, 2, and 3 are 11, 14, and 17, respectively. While using the sum of probability weighting equal to 1, the maximum amounts of targets 1, 2, and 3 are 10, 10, and 17, respectively. While using the improved algorithm, the maximum amounts of targets 1, 2, and 3 are 10, 10, and 15, respectively.

In Figs. 4.27, 4.28, 4.29, 4.30, 4.31, and 4.32, the dotted lines show the real trajectories, and solid lines show the filtering tracks.

As shown in Figs. 4.27, 4.28, and 4.29, the network radar countermeasure system uses the sum of probability weightings equal to 1. The improved algorithm combined with the uniformly accelerated motion model and the Singer model can track a target making uniform and circular motions, which enables the tracking of cross-movement targets accurately. The simple joint probability data association algorithm RMS error value is apparently bigger than the RMS error values of the other two algorithms on tracking the targets' positions. The improved algorithm has better tracking performance than the sum of probability weighting equal to 1. In this situation, it shows in the best tracking for target 1.

4.6.3.5 Simulation Example Five

During the clutter producing process, parameter $\lambda = 10^{-7}$. The rest of the simulation conditions are the same those in simulation example two. The simulation results are shown in following images:

From the simulation, while using the simple joint probability data association algorithm, the maximum amounts of targets 1, 2, and 3 are 17, 147, and 51, respectively. While using the sum of probability weighting equal to 1, the maximum amounts of targets 1, 2, and 3 are 17, 17, and 11, respectively.

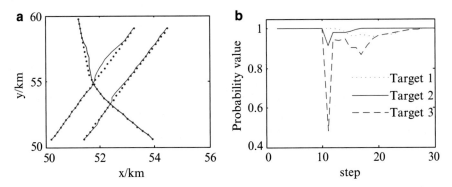

Fig. 4.27 (**a**) Simple joint probability association algorithm tracking results. (**b**) Corresponding association probabilities

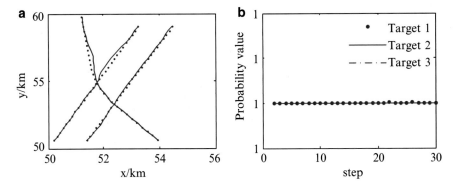

Fig. 4.28 (**a**) Association algorithm tracking results sum of weights equal to 1. (**b**) Corresponding association probabilities

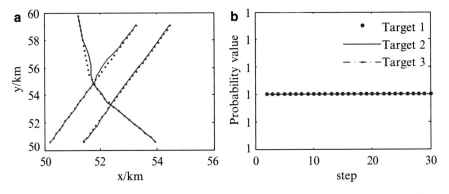

Fig. 4.29 (**a**) Improved algorithm for tracking results. (**b**) Corresponding association probabilities

Fig. 4.30 The RMS error
curve of target 1 position

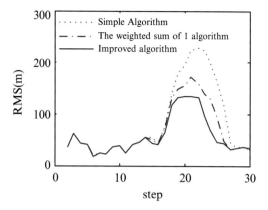

Fig. 4.31 The RMS error
curve of target 2 position

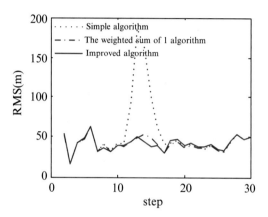

Fig. 4.32 The RMS error
curve of target 3 position

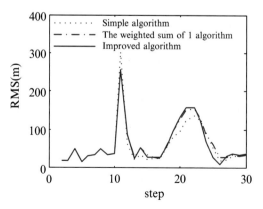

In Figs. 4.33, 4.34, 4.35, 4.36, and 4.37, the dotted lines show the real trajectories, and the solid lines show the filtering tracks.

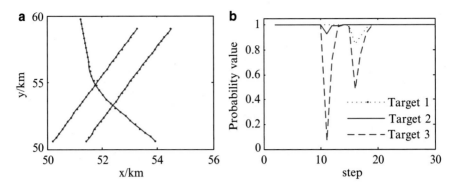

Fig. 4.33 (**a**) Simple joint probability. (**b**) Corresponding association algorithm tracking results probabilities

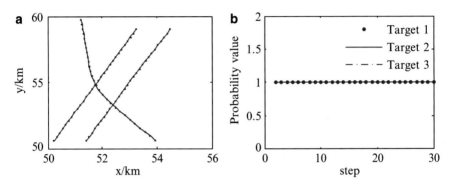

Fig. 4.34 (**a**) Association algorithm tracking results sum of weights equal to 1. (**b**) Corresponding association probabilities

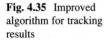

Fig. 4.35 Improved algorithm for tracking results

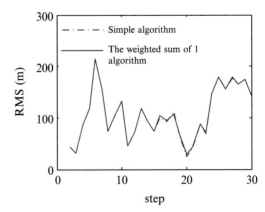

As shown in Figs. 4.33 and 4.34, the network radar countermeasure system can combine the simple joint probability data association algorithm and weights equal to 1, while the three transmitting stations work simultaneously and can track three

Fig. 4.36 Corresponding
association probabilities

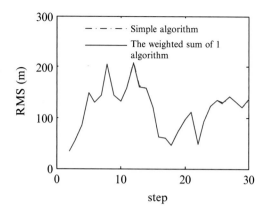

Fig. 4.37 The RMS error
curve of target 3 position

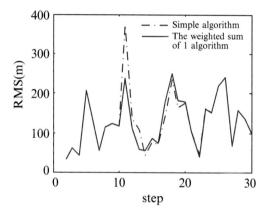

targets making cross-motions. The tracking performance of the sum of probability
weighting equal to 1 algorithm is better than the simple joint probability data
association algorithm for tracking target 3, while it is almost the same when
tracking targets 1 and 2.

Conclusion The network radar countermeasure system has completely similar
conclusions in tracking non-motor targets when using the probabilistic data association algorithm to track maneuvering targets.

4.6.3.6   Simulation Example Six

The measurement types and measurement errors are the same in every receiving
station. The transmission stations' coordinates are $T(-10, 0, 0.05)$km, the coordinates of receiving station 1 are $R_1(30, 0, 0)$km, station 2 $R_2(0, 30, 0)$km, and station
3 $R_3(0, -30, 0)$km. The measuring amount of the receiving station is the distance

sum, orientation, and pitching. using the uniform motion model. The initial values
of the three targets are:

$$X_1 = [50104, 17, 50385, 98, 10022, 0]^T,$$
$$X_2 = [51179, 0, 50379, 100, 9951, 0]^T,$$
$$X_3 = [52337, -17, 50365, 98, 9967, 0]^T,$$

The error covariance matrix of the initial state is:

$$P = \begin{bmatrix} 6400 & 1600 & 0 & 0 & 0 & 0 \\ 1600 & 800 & 0 & 0 & 0 & 0 \\ 0 & 0 & 6400 & 1600 & 0 & 0 \\ 0 & 0 & 1600 & 800 & 0 & 0 \\ 0 & 0 & 0 & 0 & 100 & 25 \\ 0 & 0 & 0 & 0 & 25 & 12.5 \end{bmatrix}$$

Because $\lambda = 4$, the rest of the simulation conditions are the same as those of
simulation example one. Using the simple joint probability data association algorithm to track the targets, the simulation results are as follows:

In Figs. 4.38, 4.39, 4.40 and 4.41, the dotted lines show the real trajectory, and
solid line shows filtering tracks. It can be seen from the Figthree, the receiving
station distance and azimuth, pitch based on simplified joint probabilistic data
association algorithm using sequential filtering structure better completion of the
three cross moving target tracking. With the number of receiving station growing
up, filter position RMS error became smaller, indicate filtering effect became better,
for target 2, using filtering effect of three receiving station is mach better than use
just one receiving station ore two. The filtering effect of target 1 and 3 is not as
obvious as target 2.

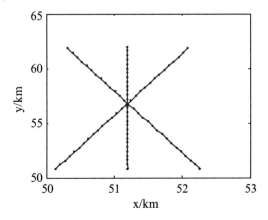

Fig. 4.38 Three receiving
stations tracking results

Fig. 4.39 The RMS error
position of target 1

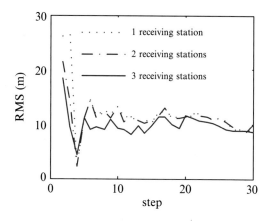

Fig. 4.40 The RMS error
position of target 2

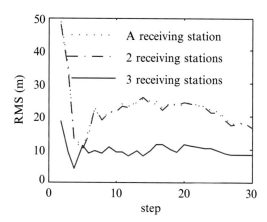

Fig. 4.41 The RMS error
position of target 3

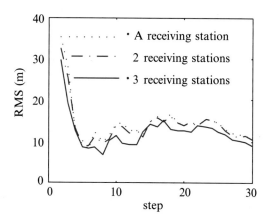

4.6.3.7   Simulation Example Seven

Coordinates of transmission station 1 $T_1(-10, 0, 0.05)$km, station 2 $T_2(10, 0, 0.05)$ km, station 3 $T_3(0, -10, 0.05)$, coordinates of receiving station 1 $R_1(30, 0, 0)$km, station 2 $R_2(0, 30, 0)$km, station 3 $R_3(0, -30, 0)$km. Measurements of every station are three range sum. The standard deviation of range sum is 3 m. The initial value of target state and its covariance matrix is same to simulation example 6, $\lambda = 2 \times 10^{-5}$. Tracking model is uniform motion. Using Simple joint probability data association algorithm to track. simulation results as follow:

In Figs. 4.42, 4.43, 4.44, and 4.45, the dotted lines show the real trajectories, and the solid lines show the filtering tracks. It can be seen that the receiving station distance, azimuth, and pitch based on the simplified joint probabilistic data association algorithm using a sequential filtering structure enabled better completion of the three cross-moving targets tracking. With the number of receiving stations increasing, the filter position RMS error became smaller, indicating better filtering effect, for target 2. Using the filtering effect of three receiving stations is much better than using just one or two receiving stations. The filtering effects of targets 1 and 3 are not as obvious as that for target 2.

Conclusion Simulation examples six and seven indicate that the network radar countermeasure systems use the measurements of every receiving station, when every station has the same measurement and same measurement error. For the multi-target data association algorithm, the filtering performance of the target improves as the number of receiving stations increases.

4.6.3.8   Simulation Example Eight

The measurement of receiving stations 1 and 2 is the range sum, but the measurements of station 3 are the range sum, azimuth, and pitch. $\lambda = 10^{-5}$. The rest of the simulation conditions are the same as those in simulation example seven. The simulation results are:

In Figs. 4.46, 4.47, 4.48, and 4.49, the dotted lines show the real trajectories, and the solid lines show the filtering tracks.

As shown in images, while the measurements of the three receiving stations are different, the use of sequential filtering was shown to be good in tracking three cross-moving targets. Every receiving station in the network radar countermeasure system uses the sequential data association algorithm to track target 3, measuring the amount of different types. With a greater number of receiving stations, the filtering performance increases, which is most obvious for target 2, whereas targets 1 and 3 did not show the same as that for target 2.

Fig. 4.42 Three receiving stations tracking results

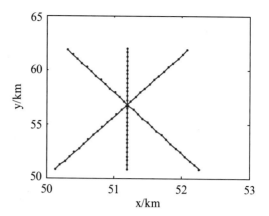

Fig. 4.43 The RMS error station position of target 1

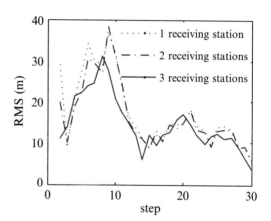

Fig. 4.44 The RMS error position of target 2

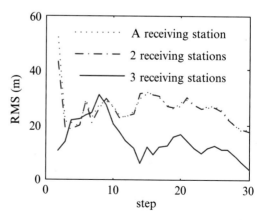

Fig. 4.45 The RMS error
position of target 3

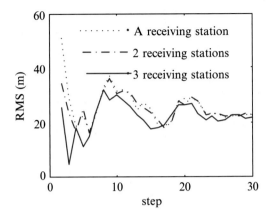

Fig. 4.46 Three receiving
stations tracking results

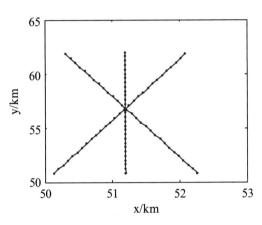

Fig. 4.47 The RMS error
position of target 1

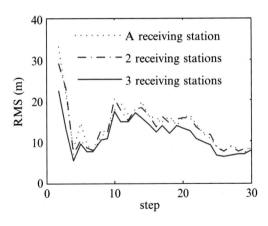

Fig. 4.48 The RMS error
position of target 2

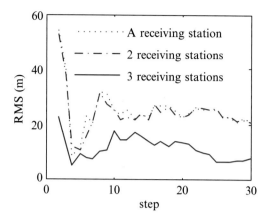

Fig. 4.49 The RMS error
position of target 3

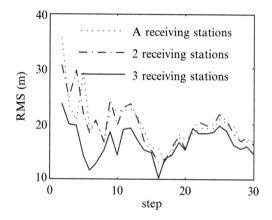

4.6.3.9   Simulation Example Nine

The type of each receiving station is the same, with different measurement error,
but the range, target's motion status, the initial value of the state of the filter, and the
initial covariance matrix are same as in simulation example seven. The measurement of every station is distance sum, the standard deviation of receiving station 1's
distance sum is 1 m, the standard deviations of receiving stations 2 and 3 are 10 m,
$\lambda = 10^{-5}$, which used in clutter production. It is a uniform motion model. The
simulation results are as follows:

In Figs. 4.50, 4.51, 4.52, and 4.53, the dotted lines show the real trajectories, and
the solid lines show the filtering tracks.

As shown in the images, in the condition of every receiving station's measurements is the same, measurement error is different, and using the ordered filter
structure can realize better tracking of the three cross-moving targets. When the
receiving station of the network radar countermeasure system has the same type of
measurement but different accuracy, the tracking effect of the receiving station with

Fig. 4.50 Three receiving
stations tracking results

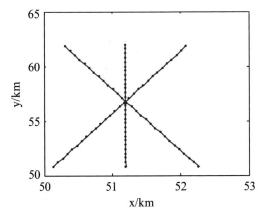

Fig. 4.51 The RMS error
position of target 1

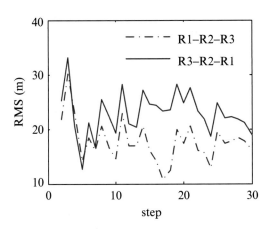

Fig. 4.52 The RMS error
position of target 2

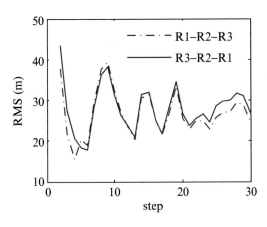

Fig. 4.53 The RMS error
position of target 3

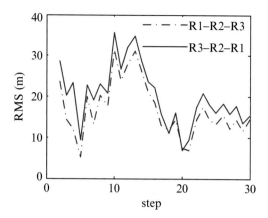

a small measurement error is better than the tracking effect of the receiving station
with a larger measurement error. The difference is obvious, and the performance in
the location RMS curve is smooth and reaches a convergence value. The reason for
this is that it is filtered based on the good performance of the receiving station, so as
to obtain more accurate estimates of the intermediate state, with lower computational complexity and smaller gates sizes. We can then deal better with the rest of
the receiving station measurements.

4.6.3.10   Simulation Example Ten

- Multiple transmission single reception (three transmitters single receiver, double
  transmitters single receiver)

(a) Double transmission and single reception
        Simulation conditions set up. the coordinates of transmission station 1 are
    $T_1(-10, 0, 0.05)$km , the coordinates of transmission station 2 are
    $T_2(10, 0, 0.05)$km, and the coordinates of transmission station 3 are R
    $(30, 0, 0)$km. The standard deviation of the target's position error is 3 m, and
    the measurement errors of the azimuth and pitch angle are both 3 mrad. The
    measurement error of the target's position is 100 m in the X and Y directions
    and 30 m in the Z direction. Target is in uniform motion, the initial values are
    $(60, 60, 10)$km, speed $v_x = 150$m/s, $v_y = -200$m/s, $v_z = 100$m/s, sampling
    period $T = 1$s, simulation time 100 s, the equation of state uses the uniform
    motion model, the state vector of the target is $X = [x, \dot{x}, y, \dot{y}, z, \dot{z}]^T$, and the
    number of Monte Carlo simulation iterations is 30. The filter's initial value is
    $X_0 = [60380, 155, 59680, -198, 10230, 102]^T$, and the covariance matrix of
    the initial state is:

Fig. 4.54 Target true track and filter values

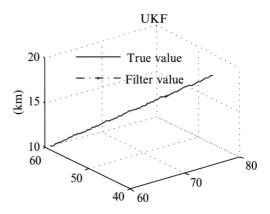

$$P_0 = \begin{bmatrix} 10000 & 10000 & 0 & 0 & 0 & 0 \\ 10000 & 20000 & 0 & 0 & 0 & 0 \\ 0 & 0 & 10000 & 10000 & 0 & 0 \\ 0 & 0 & 10000 & 20000 & 0 & 0 \\ 0 & 0 & 0 & 0 & 900 & 900 \\ 0 & 0 & 0 & 0 & 900 & 1800 \end{bmatrix},$$

The process noise covariance matrix is

$$Q = \begin{bmatrix} 0.0001 & 0.0001 & 0 & 0 & 0 & 0 \\ 0.0001 & 0.0001 & 0 & 0 & 0 & 0 \\ 0 & 0 & 0.0001 & 0.0001 & 0 & 0 \\ 0 & 0 & 0.0001 & 0.0001 & 0 & 0 \\ 0 & 0 & 0 & 0 & 0.0001 & 0.0001 \\ 0 & 0 & 0 & 0 & 0.0001 & 0.0001 \end{bmatrix},$$

We use the UKF ($\alpha = 0.01, \kappa = 0, \beta = 2$) for filtering and comparison with the EKF (Figs. 4.54, 4.55, 4.56, 4.57, and 4.58). The simulation results are as follows:

(b)   Double transmission and single reception. The measurements of the receiving station are two distance sums and an azimuth

The simulation conditions are same as for part (a) (Figs. 4.59, 4.60, and 4.61), and the simulation results are as follows:

Based on the two distance and the position RMS error value with an azimuth angle for tracking the position RMS error margin of the position based on the two distance and a pitch angle is rapidly converging to the following 50 m, and are approaching their Cramer–Rao bound. At the same time, based on the two distance and the speed of a RMS error tracking the speed of the error value is based on two distance and a pitch angle tracking speed RMS error value is also quickly converge to a smaller value. Show that in the case of double transmission and single reception, Based on measurement of two distance and a

Fig. 4.55 Location RMS error values

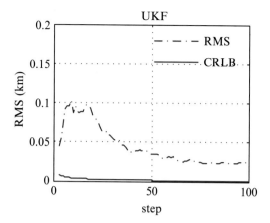

Fig. 4.56 Speed RMS error values

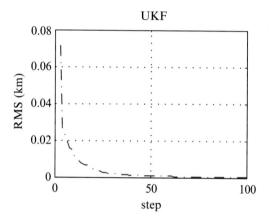

Fig. 4.57 Target true track and filter values

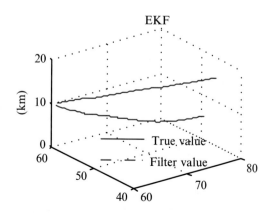

Fig. 4.58 Location RMS error values and CRLB values

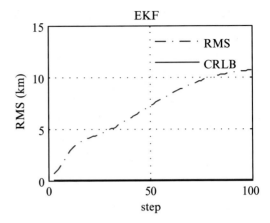

Fig. 4.59 Target true track and filter values

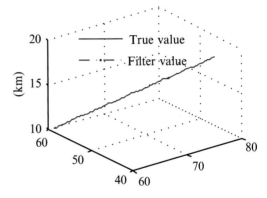

Fig. 4.60 Location RMS error values

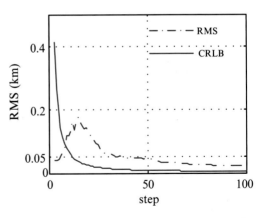

pitching angle and based on two measuring distances and a range angle can be successfully completed tracking the single moving target. Based on the measurements of two distance and a pitch angle, the position of the RMS peak was significantly less than that of the RMS peak value based on the measurements

Fig. 4.61 Speed RMS error values

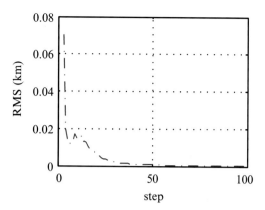

of the two distance and an azimuth tracking. And the position RMS value quicker converges to the speed of the following 50 m. Based on the measurement of the two distance and pitch angle, the tracking performance is better than the tracking performance based on the measurement of the two distance and the azimuth angle. Based on the measurement of two distance and a pitch angle combined with EKF filtering algorithm, the filtering results are divergent, This is because the EKF filtering algorithm is not suitable for dealing with the highly nonlinear of the measurement equation. In view of the disadvantages of the EKF filter algorithm, a unified UKF filtering algorithm is used in the following simulation.

As shown in Figs. 4.62, 4.63, and 4.64, in three rounds and a receive mode, with the distance and position based on three combined RMS error values, the UKF single target tracking filter algorithm quickly converges to 50 m or less, and the speed RMS error value quickly converges to an even smaller value, which shows that the mode of the three distances can accurately track the moving target. In the case of the three tranomitters work station, its location RMS error curves converge to speed 50 m and the velocity RMS error curve converges to a minimum value of speed faster than the result based on both distance and azimuth and based on both the distance and the convergence speed of a pitch angle tracking; At the same time, three conditions of the position RMS error curve and peak velocity RMS error curve less. Showed in a collection mode is based on three rounds of three distances and tracking performance was better than the next two rounds of a collection mode and with a pitch and track the performance of the two distances and with an azimuth based on two distances.

Fig. 4.62 Real track and
filter values

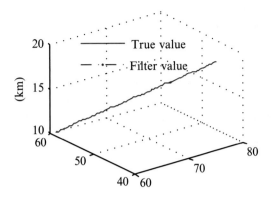

Fig. 4.63 Location RMS
error values

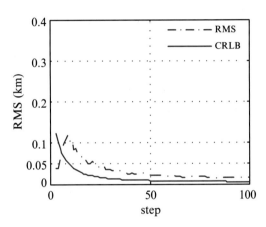

Fig. 4.64 Speed RMS error
values

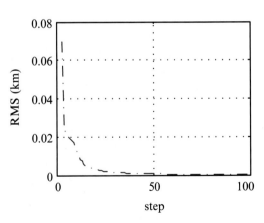

4.6.3.11 Simulation Example Eleven

Single transmitter multiple receivers (single out four reception). The simulation conditions are set as follows: the coordinates of the transmitting station T(-10, $0, 0.05$)km, the coordinates of receiving station 1 R1($30, 0, 0$)km, the coordinates of receiving station 2 R2($0, 30, 0$)km, the coordinates of receiving station 3 R3 ($0, -30, 0$)km, and the coordinates of receiving station 4 R4($-30, 0, 0$)km. The matrix of the initial interaction between the local estimation error covariance of receiving stations i and j is:

$$P^{ij}_{0/0} = \begin{bmatrix} 0 & 0 & 0 & 0 & 0 & 0 \\ 0 & 0 & 0 & 0 & 0 & 0 \\ 0 & 0 & 0 & 0 & 0 & 0 \\ 0 & 0 & 0 & 0 & 0 & 0 \\ 0 & 0 & 0 & 0 & 0 & 0 \\ 0 & 0 & 0 & 0 & 0 & 0 \end{bmatrix},$$

Process noise covariance matrix:

$$Q = \begin{bmatrix} 0.1 & 0.1 & 0 & 0 & 0 & 0 \\ 0.1 & 0.1 & 0 & 0 & 0 & 0 \\ 0 & 0 & 0.1 & 0.1 & 0 & 0 \\ 0 & 0 & 0.1 & 0.1 & 0 & 0 \\ 0 & 0 & 0 & 0 & 0.1 & 0.1 \\ 0 & 0 & 0 & 0 & 0.1 & 0.1 \end{bmatrix}.$$

The other simulation conditions are same as in simulation example one. We examine the following three conditions: (1) distributed without feedback; (2) distributed feedback; (3) under centralized tracking performance. The simulation results are shown in Figs. 4.65, 4.66, 4.67, 4.68, 4.69, 4.70, 4.71, 4.72, 4.73, and 4,74,

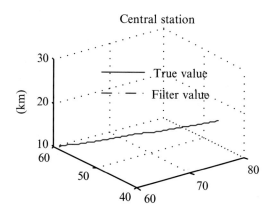

Fig. 4.65 Target track and filter real values (not distributed feedback)

Fig. 4.66 Position RMS
error values (distributed no
feedback)

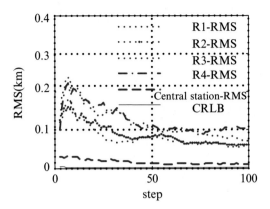

Fig. 4.67 Speed RMS error
(not distributed feedback)

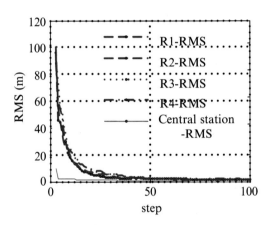

Fig. 4.68 Location RMS
error comparison (not
distributed feedback)

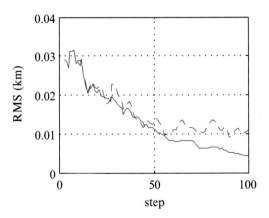

Fig. 4.69 The subject's real tracks and filtered values (distributed feedback)

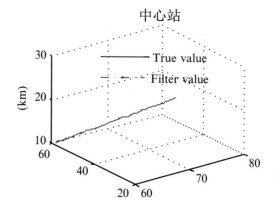

Fig. 4.70 Location RMS error values (distributed feedback)

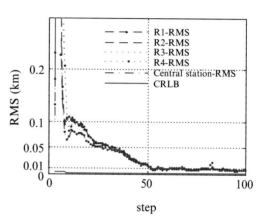

Fig. 4.71 Speed RMS error values (distributed feedback)

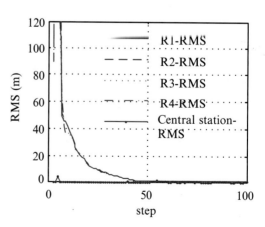

Fig. 4.72 The subject's
real tracks and filtered
values (centralized)

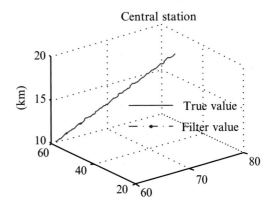

Fig. 4.73 Location RMS
error values (centralized)

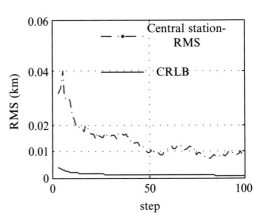

Fig. 4.74 Speed RMS error
values (centralized)

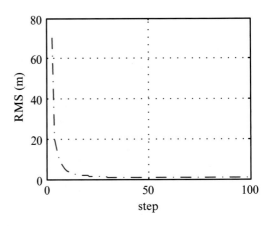

Distributed (feedback, no feedback) structure: The centralized structure could be used to complete the single tracking of moving objects. The centralized structure's tracking performance is the best, as it reaches the speed of location RMS error value 0.01 km faster. Tracking the performance of the distributed architecture with and without feedback distributed architecture fairly, the feedback did not improve the overall tracking performance. This shows that, for both the location and speed the filtered RMS error final convergence value, except that for each station, the estimation error covariance matrix is reduced, with the previous theoretical analysis. The tracking performance of the distributed central station is very close to the centralized tracking performance without feedback, and the tracking performance of the central station is much better than the tracking performance of each individual station. The RMS position error value is closer to the Cramer–Rao bound. When the process is noisy (e.g., simulation conditions setting), the state estimation calculation error covariance between the receiving stations cannot be ignore. The reason for this is that ignoring the position of the RMS error value of each station's obtained correlation value of the actual position RMS error should yield a larger difference. Only when there is small process noise, before considering the ignored state estimation calculation error between the receiving station covariance, which can improve the efficiency and does not have a big impact on the actual results.

4.6.3.12 Simulation Example Twelve

Multiple transmitters and multiple receivers (three transmit stations and three receiving stations). The simulation conditions are set up as follows: transmit station 1 coordinates T1(−10, 0, 0.05)km, transmit station 2 coordinates T2(10, 0, 0.05) km, transmit station 3 coordinates T3(0, −10, 0.05)km, receiving stations coordinates R1(30, 0, 0)km, R2(0, 30, 0)km, and R3(0, −30, 0)km. The measurement of receiving stations is the distance sum, and the other simulation conditions are same as for simulation example two. The investigation results of the distributed without feedback and centralized systems tracking performance are shown in the following figures:

In the case of multiple transmitters and multiple receivers, as shown in Figs. 4.75, 4.76, 4.77, 4.78, 4.79, and 4.80, the distributed feedback structure with no centralized structure is well suited to tracking moving objects. The RMS error performance of the position values approaching their CRB circles, and the speed RMS value quickly converges to a smaller value. The tracking performance of the distributed architecture approaches the performance of the central station centralized structure, and is significantly better than tracking the performance of each receiving station.

Fig. 4.75 Target tracks and filter real values (not distributed feedback)

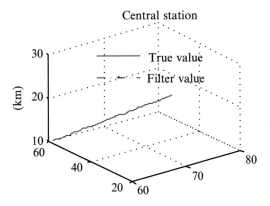

Fig. 4.76 Position RMS error values

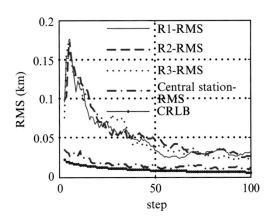

Fig. 4.77 Speed RMS error values

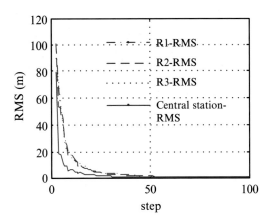

Fig. 4.78 The subject real tracks and filtered values (centralized)

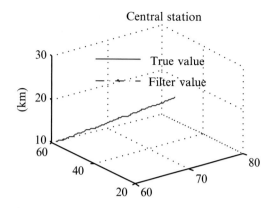

Fig. 4.79 Position RMS error values

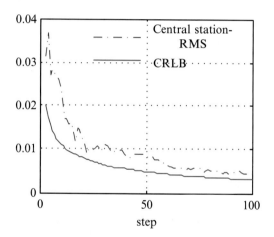

Fig. 4.80 Speed RMS error values

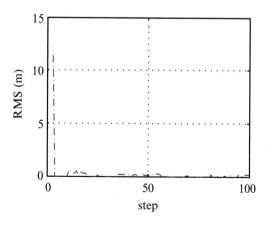

4.6.3.13   Simulation Example Thirteen

The central Internet station's coordinates are R0 = (0, 0, 0)km, and the receiving stations coordinates are R1(30, 0, 0), R2(0, −30, 0)km, and R3(−21.2, 21.2, 0)km. The difference measurement error is 3 ns, electromagnetic wave propagation speed $c = 3 \times 10^8$ m/s, the initial position of the target is (150, 150, 10)km, target movement time $t = 0 \sim 190$ s, wherein $t = 0 \sim 20$ s. The goal to perform uniform motion, with speed $v_x = 0$ m/s, $v_y = -200$ m/s, $v_z = 0$ m/s $t = 20 \sim 100$ s, carry out a slow turn, then accelerate at $a_x = 3$ m/s^2, $a_2 = 3$ m/s^2, $a_z = 0$ m/s^2, perform uniform motion in $t = 100 \sim 120$ s, $t = 120 \sim 140$ s, carry out a fast turn, accelerate at $a_x = -15$ m/s^2, $a_y = -15$ m/s^2, $a_z = 0$ m/s^2, and perform uniform motion in $t = 140 \sim 190$ s. The initial value of the state of the filter is $X_0 = [150060, 1, 150280, -205, 10080, 1]^T$, and the initial covariance matrix is:

$$
P_{0/0} = \begin{bmatrix}
10000 & 10000 & 0 & 0 & 0 & 0 \\
10000 & 20000 & 0 & 0 & 0 & 0 \\
0 & 0 & 10000 & 10000 & 0 & 0 \\
0 & 0 & 10000 & 20000 & 0 & 0 \\
0 & 0 & 0 & 0 & 900 & 900 \\
0 & 0 & 0 & 0 & 900 & 1800
\end{bmatrix},
$$

Using the IMM algorithm for target tracking, the motion model uses the uniform motion and constant acceleration model. The initial model probability is (1, 0, 0), and the model transition probability matrix is:

$$
P = \begin{bmatrix}
0.95 & 0.025 & 0.025 \\
0.025 & 0.95 & 0.025 \\
0.025 & 0.025 & 0.95
\end{bmatrix},
$$

Model 1 (uniform motion model) process noise covariance matrix:

$$
Q_1 = \begin{bmatrix}
10^{-8} & 10^{-8} & 0 & 0 & 0 & 0 \\
10^{-8} & 10^{-8} & 0 & 0 & 0 & 0 \\
0 & 0 & 10^{-8} & 10^{-8} & 0 & 0 \\
0 & 0 & 10^{-8} & 10^{-8} & 0 & 0 \\
0 & 0 & 0 & 0 & 10^{-8} & 10^{-8} \\
0 & 0 & 0 & 0 & 10^{-8} & 10^{-8}
\end{bmatrix},
$$

Model 2 (uniformly accelerated motion model) process noise covariance matrix:

$$Q_2 = \begin{bmatrix} 0.000625 & 0.00125 & 0 & 0 & 0 & 0 & 0.00125 & 0 & 0 \\ 0.00125 & 0.0025 & 0 & 0 & 0 & 0 & 0.0025 & 0 & 0 \\ 0 & 0 & 0.000625 & 0.00125 & 0 & 0 & 0 & 0.00125 & 0 \\ 0 & 0 & 0.00125 & 0.0025 & 0 & 0 & 0 & 0.0025 & 0 \\ 0 & 0 & 0 & 0 & 0.00125 & 0.0025 & 0 & 0 & 0.0025 \\ 0 & 0 & 0 & 0 & 0.00125 & 0.0025 & 0 & 0 & 0.0025 \\ 0.00125 & 0.0025 & 0 & 0 & 0 & 0 & 0.0025 & 0 & 0 \\ 0 & 0 & 0.00125 & 0.0025 & 0 & 0 & 0 & 0.0025 & 0 \\ 0 & 0 & 0 & 0 & 0.00125 & 0.0025 & 0 & 0 & 0.0025 \end{bmatrix},$$

Model 3 (uniformly accelerated motion model) process noise covariance matrix:

$$Q_3 = \begin{bmatrix} 0.14063 & 0.28125 & 0 & 0 & 0 & 0 & 0.28125 & 0 & 0 \\ 0.28125 & 0.5625 & 0 & 0 & 0 & 0 & 0.5625 & 0 & 0 \\ 0 & 0 & 0.14063 & 0.28125 & 0 & 0 & 0 & 0.28125 & 0 \\ 0 & 0 & 0.28125 & 0.5625 & 0 & 0 & 0 & 0.5625 & 0 \\ 0 & 0 & 0 & 0 & 0.14063 & 0.28125 & 0 & 0 & 0.281250 \\ 0 & 0 & 0 & 0 & 0.28125 & 0.5625 & 0 & 0 & 0.5625 \\ 0.28125 & 0.5625 & 0 & 0 & 0 & 0 & 0.5625 & 0 & 0 \\ 0 & 0 & 0.28125 & 0.5625 & 0 & 0 & 0 & 0.5625 & 0 \\ 0 & 0 & 0 & 0 & 0.28125 & 0.5625 & 0 & 0 & 0.5625 \end{bmatrix},$$

The results after 30 iterations of the Monte Carlo simulation are shown in Figs. 4.81, 4.82, 4.83, 4.84, 4.85, 4.86, 4.87, 4.88, 4.89, 4.90, and 4.91.

The algorithm is based on the position, velocity, and acceleration estimated RMS error value. As shown, the position can be obtained about the peak RMS error 300 m, the smallest value is approximately 2 m, and the RMS error rates from the obtained acceleration peak and peak RMS are 100 m and 22 m, while the minimum

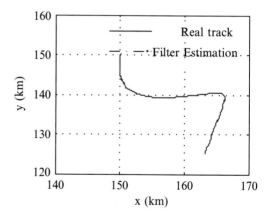

Fig. 4.81 The subject real tracks and filter values

Fig. 4.82 Target speed and the true filter values

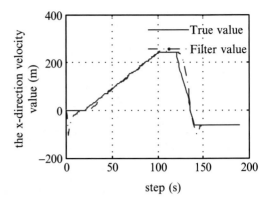

Fig. 4.83 Target speed and the true filter values

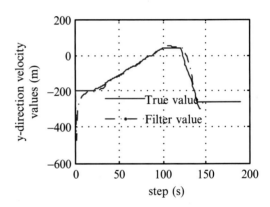

Fig. 4.84 Target speed and the true filter values

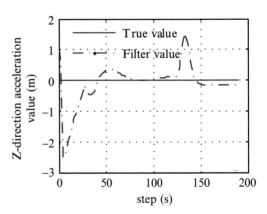

Fig. 4.85 Target
acceleration value and the
true filter values

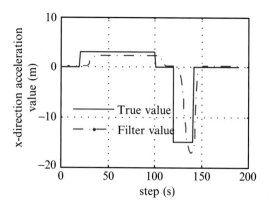

Fig. 4.86 Target
acceleration value and the
true filter values

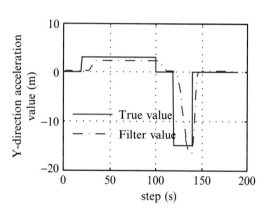

Fig. 4.87 Target
acceleration value and the
true filter values

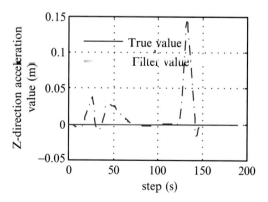

Fig. 4.88 Probability
values of the model at
different times

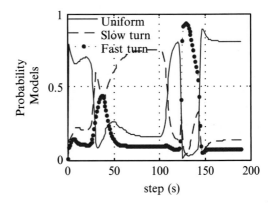

Fig. 4.89 Position RMS
error values and CRLB
values

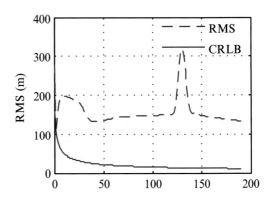

Fig. 4.90 Speed RMS error
values

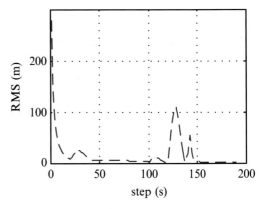

Fig. 4.91 Acceleration
RMS error values

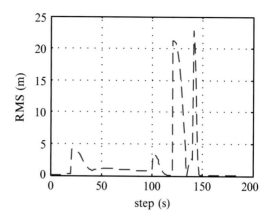

is about 2 m. Speaking from the perspective of the error algorithm, error is feasible. The effectiveness of the IMM tracking algorithm for the maneuvering target is demonstrated. The figure also shows the probability of the models at different times. As $p = 0.5$, a demarcation point conversion model can be seen from the figure the moment at which the target maneuvering occurred.

Chapter 5
Network Radar Countermeasure Systems

5.1 Introduction

The resource and data information of network radar countermeasure systems are widely shared via the network, which not only brings great convenience and flexibility of processing in target identification and target route connection, but also improves processing in terms of accuracy and real-time application. The target track route connection is an important expression of superiority of an investigation probe of integration in active and passive modes, and is also an important function of the system.

The central station makes a judgment based on every piece of local information and target message, to decide whether every piece of local information corresponds to the same target. Every local station provides the target's position and identification message to the central station. The target's position message means that dynamic parameters are used to describe the target's state of motion; for example, the position and speed, etc. The receiving station also obtains the identification message of the enemy target by passive detection, so the target's identification message corresponds to every local flight route.

In traditional electronic-intelligence-reconnaissance system, because the active detection information and passive investigation information cannot be shared in real time, the identification information of passive investigation cannot be integrated into the target route information gained by active detection. The network radar countermeasure system integrates the information of both active detection and passive investigation, and can share the target's information in real time at the receiving station, so we can use the targets' identification information as a basis for data association.

The principle of target connection and track fusion at the central station is shown in Fig. 5.1.

© National Defense Industry Press, Beijing and Springer-Verlag Berlin Heidelberg 2016   243
Q. Jiang, *Network Radar Countermeasure Systems*,
DOI 10.1007/978-3-662-48471-5_5

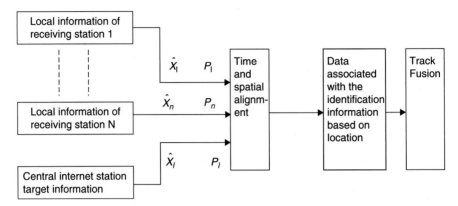

Fig. 5.1 The principle of target connection with target route at the central station

5.2 The Pretreatment of a Network Radar Countermeasure System

The central station integrates every local station's information, and completes the integration of the information about the target. But before integration of the data, the central station needs to change the coordinates of every local station's data. At the same time, because the observation data are not synchronized with every local receiving station in terms of observing the target, recording data, communicating, and central data processing, the central station has to make an alignment of time and space, and that is the objective of data pretreatment.

The experiment proves that the receiving station has one more target route because of the existence of system deviation. If estimation and correction of the system deviation is not carried out, it will lead to failure to finish fusion of the data, which will reduce the accuracy.

5.2.1 The Space Calibration Network Radar Countermeasure System

One important problem which the central station faces is the space alignment among observation stations on the Earth at one particular time. It is always easy for engineering treatment in the past to ignore the impact of the curvature of the surface of an ellipsoid. A space calibration method of low accuracy method would be used, which fails to fuse the route data of the same object, reducing the accuracy of fusion, and increasing the errors in the flight route.

The Earth is a giant ellipsoid, and low-altitude flights actually allow the surface of the ellipsoid to move. Therefore, different Earth models in different applications should be used such as: plane model, round ball model, and ellipsoid model (Fig. 5.2).

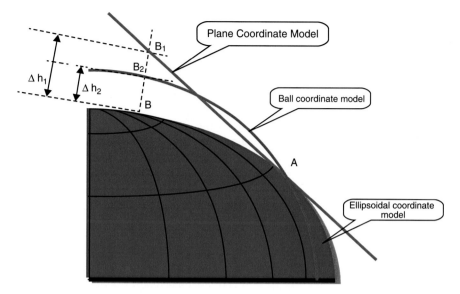

Fig. 5.2 Three kinds of different schematic models of the Earth

The radius of the Earth is very large, which causes only a small curvature at a single point. So, it is easy to ignore the bending at the ground level. When the movement and observation zone are contained within a small area, all the same sea-level points can be regarded as being in an ideal plane. The plane model is the simplest of the three models, but it also has the lowest accuracy. When the plane model is used in multi-sensor data fusion, the changing of coordinates between every sensor's local coordinate system is a panning relationship.

Through experimentation, the oblations of Earth are $f = (a-b)/a$ too small, which means that the Earth's half shaft is much similar and it is close to a spheroplast, and, in some applications, the Earth is described as a round ball model. In a round ball model, the center of the ball and earth's mass coincide at one point. The radius uses a middle value between the long and short axes, which, according to the difference in latitude of specific positions, will change. In terms of accuracy, the sphere model is accurate in longitude, but has a great error in latitude. The sphere model is used in multi-receiving-station data fusion, and the change of coordinates between every sensor's local coordinate system is a panning and rotation relationship. Because the oblateness is ignored, it's easy to calculate the spin matrix.

The ellipsoid model considers the Earth as an ellipsoid, where the major axis is a, the minor axis is b, and all of the wefts are circles. The equatorial round radius is a and the Earth's axis is b. The ellipsoid model has very high accuracy as well as a complex design.

The local coordinate system is the receiving station's own probe coordinate system. An important feature of this coordinate system is to set a specific point as the center of the coordinate system, and use the geoids of this coordinate system center to build a polar spherical coordinate system. It involves slant range r, azimuth θ, and pitch angle ϕ.

Local coordinates can reflect the coordinates of somewhere inside of the space efficiently. However, to coordinate the position in a wider range of areas, it needs latitude and longitude coordinates or an Earth geocentric rectangular coordinate system.

An Earth geocentric rectangular coordinate system, which uses the Earth's geocentric as the origin of the coordinate system, has the axis of rotation as the Z axis, positive direction as north, positive direction of the X axis as the crossover point with the equatorial plane, positive direction of the Y axis through the equatorial plane and the crossover point with the equatorial plane makes a 90° angle. Earth geocentric coordinates are not visual and cannot provide geographical location information, but it is a necessary middle coordinate system in changing every kind of coordinate.

The latitude and longitude coordinates system uses longitude, latitude, and altitude to indicate specific geographic positions. It uses the Earth's center as the coordinate system's center. The definition of longitude considers the Royal Observatory in Greenwich as 0°. Around the Earth's axis from west to east it turns in several angles. At any point between the latitude defined as the ellipsoid normal and the equatorial plane of the point of the angle, it is defined as positive in the northern hemisphere, and the southern hemisphere is defined as negative. Altitude is defined as a point to point range ellipsoid; that above the Earth's ellipsoid is positive, otherwise it is negative. It can be used to calculate the latitude and longitude of the Earth ellipsoid parameters in different applications, and it may suit different ellipsoid parameters; for example, the GPS positioning system uses the WGS84 ellipsoid parameters used in the international arena, and our widely used Beijing and Xi'an contain 54 and 84 parameters, respectively.

There is a direct link between the level surface to a point on the Earth's latitude and longitude to that point. So, from the latitude and longitude of a point, a rotation matrix can be drawn between the coordinates of the center point of the local Cartesian coordinate system and the Earth, and the Earth's Cartesian coordinates position of the point is the local coordinates and the Cartesian coordinates conversion of the Earth requires a translation vector. It can be seen any local coordinate system can be converted to the Earth's rectangular coordinates, thereby switching between any two local coordinate systems can also be completed accurately.

In the central station data processing, it is required to provide each station's measured values expressed in the same coordinate system. During the distribution of the Earth's surface, each station has different longitude and latitude measurements, and as a result of the height of the individual sites, they obtain the slant range, azimuth, and pitch angle in the local coordinate system. For relevant and accurate positioning, tracking, and matching, individually measured values obtained by different sites are required, all converted into a unified coordinate system. Since it is non-parallel around the horizontal, between each local

Fig. 5.3 Local coordinate system with the Earth geocentric Cartesian coordinate system

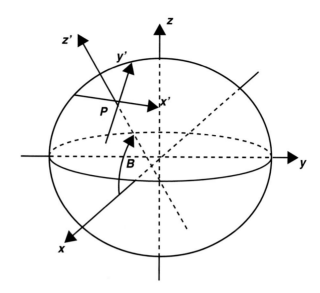

coordinate system is a certain mutual angle, which is not easy. It is easier to establish direct contact between the local coordinate system between the Earth and the geocentric Cartesian coordinate system, so the Earth geocentric Cartesian coordinate system can be used for the conversion between the local coordinate systems of intermediate coordinate systems (Fig. 5.3).

If a receiving station is located at point $p(x_0, y_0, z_0)$, it has corresponding geographic longitude, latitude, and altitude, respectively, (l, b, h). Centered on the establishment of the station as a local coordinate system, it has three axes x, y, z, as well as a unit vector corresponding to $(1 \quad 0 \quad 0)^T, (0 \quad 1 \quad 0)^T, (0 \quad 0 \quad 1)^T$. The three units provided correspond to the Earth's vector conversion to the Cartesian coordinate system. By definition, there are three axes:

$$\begin{cases} x' = (-\sin l \quad \cos l \quad 0)^T \\ y' = z' \times x' \\ z' = (\cos b \cos l \quad \cos b \sin l \quad \sin b)^T \end{cases} \tag{5.1}$$

Suppose that the local coordinate system to the Earth's coordinate system rotation matrix is R, where:

$$x' = R \cdot (1 \quad 0 \quad 0)^T, y' = R \cdot (0 \quad 1 \quad 0)^T, z' = R \cdot (0 \quad 0 \quad 1)^T \tag{5.2}$$

Therefore:

$$R = \left(x', y', z' \right)^T \tag{5.3}$$

So the Earth is at any point in a Cartesian coordinate system p, with vector v and covariance matrix P. The corresponding point of the local coordinate system is p', with vector v' and covariance matrix P'. Between the transformation formulas, we have:

$$\begin{aligned}
v &= R \cdot v' \\
v' &= R^{-1} \cdot v \\
p &= R \cdot p' + (x_0, y_0, z_0)^T \\
p' &= R^{-1} \cdot \left(p - (x_0, y_0, z_0)^T \right) \\
P &= R \cdot P' \cdot R^T \\
P' &= R^T \cdot P \cdot R
\end{aligned}$$

(5.4)

According to the above formula, the local coordinate system transformation between any two network radar countermeasure systems on different sites can be converted into the Earth's coordinate system to complete a Cartesian coordinate system.

5.2.2 Time Calibration of the Network Radar Countermeasure System

In a network radar countermeasure system, each station implements its own independent asynchronous scanning cycle work for the same target. The moment a station scouts is different, then the location information of the target is not the same. Such data cannot be directly fused on the local track data without time alignment.

In order to achieve time synchronization of the receiving station, interpolation and extrapolation of the time alignment method within the same time slice of each receiving station gathers the target observation data of extrapolation, as well as the time of high-accuracy observation data projections to the low precision of the observation time.

The steps of this method are as follows:

1. Select the time slice, and divide it with the specific moving target. The status of the target can be stationary. For low-speed and high-speed motions, the corresponding fusion time slice can be selected as hours, minutes, and seconds.
2. The observation data are collected incrementally for each receiving station according to the measurement accuracy required.
3. In order to form a series of equally spaced target observations, high precision observation data time points are obtained from low precision interpolation and extrapolation. The same time slice observation data usually have multiples, as shown in the Fig. 5.4

Fig. 5.4 Time alignment legend

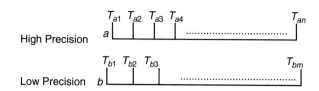

High Precision

Low Precision

There are a lot of data in each fusion time slice, which are randomly distributed in different time points. We will double the maximum error value of the receiving station terminal arrival time measurements as a normalized interval. The data collected in this period of time are considered to be at the same time, as shown in Fig. 5.4. In this case, the entire equivalent time slice is divided into several normalization intervals:

$$T_t = (T_{t+1} + T_{t+2} + T_{t+3} + \cdots T_{t+m+1})/m \tag{5.5}$$

In the above, $T_{t+1}, T_{t+2}, T_{t+3} \cdots T_{t+m+1}$ are the received data in time period T_M. These data are in uniform linear motion following interpolation or extrapolation to time T_t. Suppose the receiving station collected a vector space track in at time T_1, which gives $(X_1, Y_1, Z_1, V_x, V_y, V_z, T_1)$ after coordinate transformation, then convert to the time T_2 as:

$$\begin{aligned}
X_t &= X_1 + (T_t - T_1)*V_x \\
Y_t &= Y_1 + (T_t - T_1)*V_y \\
Z_t &= Z_1 + (T_t - T_1)*V_z \\
V_{xt} &= V_x \\
V_{yt} &= V_y \\
V_{zt} &= V_z
\end{aligned} \tag{5.6}$$

This method can be calculated by data on the high-precision observation time on the low-precision observation time to achieve synchronization of the network radar system observational data against time.

5.3 Hubs Associated Target Track

The network radar countermeasure system detects a wide range of areas within the system, and a complex, multi-objective circumstance is likely to cause a large number of false tracks. Therefore, track correlation is difficult, especially the data associated with a large time requirement for the receiving station than normal, with conversion work and a lot of uncertainties associated with the tracking process. Because of ambiguity and track association algorithm processing for a long period of time, and other reasons, the indicators to assess the value of the track correlation factor set, the actual application must be as small as possible in order to reduce the number of associations and, therefore, make full use of each station in real time, reliably sharing goal advantages and identity information, based on the target track correlating identity information, giving a better overall solution.

5.3.1 Identity Information Associated with the Target Track

The track information of each station not only provides both hubs containing the target location information, but it also contains the target identity information. Because the target identity is relatively fixed, this means that, if we can obtain the target identity information to track association, it can largely reduce the uncertainty associated with the track.

From identification friend or foe (IFF) and the passive reconnaissance identification function, the target identity can be divided into target attribute, target species, and target individual information, as shown in Fig. 5.5. Due to various reasons, the target level for each station to obtain identity information may be different, and some individual stations can obtain the target identity information, some can obtain the attribute information of the target station, and others may not obtain the target identity information, only the location information of the target's central station transmission. The consistency of the status information for each target is difficult to compare. If it is used directly for track association, a correlation algorithm may lead to an overly complex situation.

Therefore, the proposed track association method is based on objective identifiable information, and the use of different receiving stations to obtain identity information of the target tracks at different levels were associated. When the target recognition results are at the same level, the use of consistent identity information on the identity of the target is compared; if the recognition results are at different levels of information, then we go back to the same levels for comparison. When one party tracks information without identity information, we do not correlate the identity information based on the target track but track the position from directly

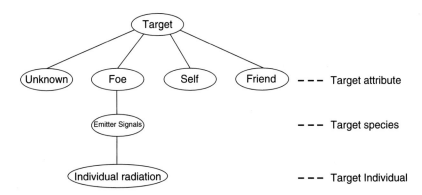

Fig. 5.5 Identity of the target

related information. For simplicity, we consider only the case of two receiving stations; multiple receiving stations follows the same approach.

Let K be time, then observation station S produced track I. This is expressed as:

$$X_i^s(k), \quad i = 1, 2, \cdots; \quad s = A, B; \tag{5.7}$$

The target property code can be expressed as M_i^{s1}, which occurs in the ranges as shown in Table 5.1.

In the command and decision-making process, the attribute information of the target is very important. If a track is given a clear target property, then the probability of a correct recognition result is relatively large. Given in Table 5.1 are the unknown target attribute and the target attribute to the enemy, friendly with a strong compatibility. Therefore, the target attribute information-related criteria can be set to:

$$M_i^{A1} = M_j^{B1} \operatorname{or} M_i^{A1} = 1 \operatorname{or} M_j^{B1} = 1 \tag{5.8}$$

It is determined the $X_i^A(k)\, X_j^B(k)$ candidates on the track. The next step may be the target type, related to the information of the target species.

If two receiving stations have completed target identification by using the target type to associate the track, we can obtain a better association effect. The target species code of track $X_i^s(k)$ can be expressed as M_i^{s2}. If there is a multi-function fire control radar, navigation radar, weather radar, imaging radar, and airborne early warning radar, the value of the reference range is in the target recognition framework; refer to Table 5.2.

Set the recognition of track $X_i^A(k)$ results of radiation source species for:

$$P_i^{A2}(m) = P\{M_i^{A2} = m\} \tag{5.9}$$

Set the recognition of track $X_j^B(k)$ results of radiation source species for:

$$P_j^{B2}(m) = P\{M_j^{B2} = m\} \tag{5.10}$$

Table 5.1 Target property code

Target attribute information	Unknown	Enemy	Ours	Friendly
Info code (M_1)	1	2	3	4

Table 5.2 Target category code

Target species information	Multi-function attack fire control radar,	Navigation radar	Weather radar,	Imaging radar	Early warning radar	...
Info code (M_2)	1	2	3	4	5	.

At this time, the target types are the same as the probability of the two tracks:

$$\eta_{ij} = \sum_m P_i^{A2}(m)P_j^{B2}(m) \tag{5.11}$$

The criterion of the related decision can be set:

$$\eta_{ij} > \eta \tag{5.12}$$

In the two successful track correlations, the threshold in the judgment formula can be determined based on simulation.

If the two tracks are given in the individual target recognition results, we can directly use the target individual identification information to associate with the target track. The track $X_i^s(k)$ target's individual code can be expressed as M_i^{s3}. The value of the reference range is the target individual recognition framework.

Setting the target individual identification of track $X_i^A(k)$ results in:

$$P_i^{A3}(m) = P\{M_i^{A3} = m\} \tag{5.13}$$

Setting the target individual identification of track $X_j^B(k)$ results in:

$$P_j^{B3}(m) = P\{M_j^{B3} = m\}, \tag{5.14}$$

At this time, the target types are the same with the probability of the two tracks:

$$\eta_{ij} = \sum_m P_i^{A3}(m)P_j^{B3}(m) \tag{5.15}$$

The criterion of the related decision can be set:

$$\eta_{ij} > \eta \tag{5.16}$$

If the two tracks carry the target individual information verdict which is consistent, it indicates that the two receiving stations' calculated trajectory is the same target, which can be directly identified that the two tracks are associated and skip the track correlation on location-based information directly, to move onto the next track fusion.

If a track gives the type of target information, another track gives the target individual information, as they do not belong to the same level and cannot be directly fused. Individual information will be traced back to the type information to the fusion type information.

Setting the target type identification of track $X_i^A(k)$ results in:

$$P_i^{A2}(m) = P\{M_i^{A2} = m\} \tag{5.17}$$

Setting the target individual identification of track $X_j^B(k)$ results in $P_j^{B3}(m)$ being converted as $P_j^{B3}(m)$ to the target type recognition level. If $P_j^{B3}(m) = P_j^{B2}(m)$ is advisable, then use the target type of information to reach an association decision.

Among two tracks, there is at least one of them that does not give information about the target. At this time, the information is insufficient, so the track cannot be associated with the decision property of the target and type information, which should be preserved before moving to the next track correlation on the location information.

In the process of track correlation with effective use of target identity information, we can reduce the complexity of the association algorithm, speeding up the processing of the track correlation. The target identity information obtained by different receiving stations is not affected by the curvature of the Earth and time errors; therefore, the use of the target identity information for association can improve, to a certain extent, the effectiveness of the track correlation algorithm, improving the overall performance of the system.

5.3.2 *Fuzzy Comprehensive Decision of Track Correlation*

The biggest disadvantage of the fuzzy track association algorithm is that the system parameter settings are very complex. For example, some parameters in the membership function are determined by a large number of simulations, with the parameters related to a set threshold. The realization of such a project is cumbersome. But in the network radar countermeasure system, since the parameters of each station are relatively uniformly fixed, the model parameters are basically consistent, and, by chance, can remedy the defects of fuzzy algorithms.

To facilitate the discussion issues, the state estimation is provided to the hubs:

$$\hat{X}_j^i (i = 1, 2, \cdots N; \quad j = 1, 2, \cdots n_j)$$

After the completion of time and spatial alignment, N is the number of receiving stations and n_j is the track number of station i. In addition, it is assumed that the data transmission delay time is zero. Unification of coordinates and time has been completed. To further simplify the analysis, this expression assumes that $N = 2$. For $N > 2$, topical processing occurs for each station, but the process does not affect the overall algorithm. Association in the process of target tracking and state estimation is not affected by the error. Suppose time t, for track i of station 1 and for track j of station 2. First, according to the state vector estimate $\hat{X}_i(l|l), \hat{X}_j(l|l)$ set up fuzzy factors between the tracks: $U = \{u_1, u_2, \cdots u_k, \cdots u_n\}$. In this formula, u_k indicates that k fuzzy factors contribute to the decision. The fuzzy factors that impact on track correlation include the target position, speed, course, acceleration, and change rate, of course. In practical applications, selecting the appropriate factors plays an important role in the association, and it can guarantee accurate tracking of the target for all kinds of movements, which does not make the algorithm too complicated and will not increase the burden of the fusion system. Select the target location (including the x, y, and z directions) and the Euclidean distance information between the target's speed as the main body of the fuzzy decision.   Calculate $\mu_k(k = 1, 2, \cdots n)$.   Set $\hat{X}_j(l|l) = \left[\hat{x}(l), \hat{y}(l), \hat{z}(l), \dot{\hat{x}}(l), \dot{\hat{y}}(l), \dot{\hat{z}}(l)\right]$. Then for the state estimation, it is desirable that $n = 2$, where:

$$
\begin{cases}
\mu_1(l) = \left[\left(\dot{\hat{x}}_i(l) - \dot{\hat{x}}_j(l)\right)^2 + \left(\dot{\hat{y}}_i(l) - \dot{\hat{y}}_j(l)\right)^2 + \left(\dot{\hat{z}}_i(l) - \dot{\hat{z}}_j(l)\right)^2\right]^{1/2} \\
\mu_2(l) = \left|\left[\left(\dot{\hat{x}}_i(l)\right)^2 + \left(\dot{\hat{y}}_i(l)\right)^2 + \left(\dot{\hat{z}}_i(l)\right)^2\right]^{1/2} - \left[\left(\dot{\hat{x}}_j(l)\right)^2 + \left(\dot{\hat{y}}_j(l)\right)^2 + \left(\dot{\hat{z}}_j(l)\right)^2\right]^{1/2}\right|
\end{cases}
$$

$$(5.18)$$

If the correlation of the two tracks' results are divided into m levels, the collection constituted by these results is called the evaluation set, denoted:

$$V = \{v_1, v_2, \cdots v_m\} \tag{5.19}$$

In the above, $v_l, l = 1, 2, \cdots m$ represents the l levels of the evaluation results. For the track valuation results of any two tracks, actually it is a fuzzy subset on V. In evaluating the association, only the target track association is of interest. So, considering the practical application and simplification of the problem, select the level of the evaluation set $m = 3$, where v_1 represents association, v_2 did not associate, and v_3 cannot be determined.

$U \times V$: From U to V, the single factor in the direct product of the fuzzy evaluation matrix set definition is:

$$R = (r_{kl})_{n \times m} \tag{5.20}$$

where r_{kl} represents k factors when considering two possible tracks to obtain first degree related results. According to the characteristics of track association fuzzy factors, the available membership type can have a normal distribution, a Cauchy type distribution, distribution center type, and we choose the normal membership function. When $m = 3$, then evaluating the level of the judgment based on a similar track, the two factors of normal MFs are:

$$r_{kl} = \exp\left(-\tau_k\left(u_k^2/\sigma_k^2\right)\right) \qquad k = 1,2 \qquad l = 1,2,3 \tag{5.21}$$

where σ_k is the fuzzy set corresponding to the k factors of the exhibition degree and τ_k is used to adjust the degree. Their value is determined by simulation.

Thus, a single factor fuzzy evaluation matrix is as follows:

$$R = \begin{bmatrix} r_{11}, r_{12}, r_{13} \\ r_{21}, r_{22}, r_{23} \end{bmatrix} \tag{5.22}$$

As the measurement accuracy of the network of radar stations against the system at a distance and azimuth varies, the weight factors need to be taken into account, wherein the weight factor fuzzy set is:

$$A = (a_1, a_2, \cdots a_n) \tag{5.23}$$

The first a_k factor is the weight coefficient corresponding to the general provisions $\sum_{k=1}^{n} a_k = 1$, and the choice of factors a_k need first degree of importance or be able influence the final decision making. Generally, choose $a_1 \geq a_2 \geq \cdots \geq a_n$ and the last weights of several factors are small. The weight coefficients of factors can be adjusted by a large number of experimental statistics. In the process of the associating network radar countermeasure system tracking, select the Euclidean distance between the target position and velocity information between the fuzzy factors, taking as precedence the positional accuracy of the network radar station against the system rather than the speed of high precision, select $a_1 = 0.6$ and $a_2 = 0.4$.

The two tracks' integrated decision-making is a composite effect of factor weights fuzzy sets A and fuzzy evaluation matrix R, namely, synthesis operations. They are synthesized by a fuzzy set on V for the two tracks on the degree of association as B:

$$B = A \cdot R = (b_l) = (b_1, b_2, b_3) \tag{5.24}$$

In the above, $b_l(l = 1, 2, 3)$ indicates the associated evaluation of the two tracks for the first two words of the membership.

A comprehensive fuzzy decision model is used mostly by the Zadeh arithmetic operation, namely:

$$b_l = \bigcup_{k=1} (a_k \wedge r_{kl}) \qquad \text{or} \qquad b_l = \bigcup_{k=1} (a_k \bullet r_{kl}) \tag{5.25}$$

As can be seen from the above two equations, because $a_k \in U$ on the weigh of the re-evaluation factors, $r_{kl} \in V$ is the evaluation level and they do not belong to a set range of values. They are grouped together, and doing it does not make sense to take a small operation. Further, when $a_k \wedge r_{kl} \leq r_{kl}$, the weight of each factor is more often correspondingly reduced, while the single factor evaluation obtained information to take small is likely to be lost, reserved only to meet $r_{kl} \leq a_k$; when there are fewer a_k, there are often correspondingly larger factors, and $a_k \wedge r_{kl} \leq a_k$ by taking small to lose all a_k greater than r_{kl}, the main factors Affecting things is always to show by the biggest a_k, so it will lose the main factors affecting the comprehensive evaluation.

The approach $a_k \cdot r_{kl}$ is part of a weighted sum. It represents the sum of the product of many factors in the conditions of membership weights and degrees of membership. It reflects that the weighted average algorithm is more in line with the connotation of fuzzy comprehensive evaluation, and, therefore, has a more universal sense of operation, so take:

$$b_l = \bigcup_{k=l}^{n} (a_k \cdot r_{kl}), \qquad l = 1, 2, 3 \tag{5.26}$$

If the decision based on the maximum membership principle degree of two track correlations is l levels, the result is that:

$$b_l = \max(b_1, b_2, b_3) \tag{5.27}$$

However, in some special cases, the reliability of the verdict of two track correlation degrees is poor. According to this maximum membership degree principle, we use the following rules for the decision making:

1. If $b_1 - b_2 > \varepsilon_1$, $b_1 > b_3$ is estimated by the track correlation;
2. If $b_2 - b_1 > \varepsilon_1$ and $b_2 \geq b_3$, it is determined that the track is not relevant;
3. The rest of the cases are determined as undefined tracks.

Among them, ε_1 is the preset threshold.

In order to associate track correlation with the history, we need to define the correlation and output quality. If the track is associated with the quality of the $C_{ra}(l)$ it is not also with the $D_{ra}(l)$ representation from the quality of the $C_{ra}(0) = D_{ra}(0) = 0$ initial value.

The one time the judgment is

$$C_{ra}(l) = C_{ra}(l - 1) + 1 \tag{5.28}$$

the one time the judgment is

$$D_{ra}(l) = D_{ra}(l - 1) + 1 \tag{5.29}$$

In order to solve a dense target environment and/or the branch and motor cross track with greater track correlation, taking two positive integers I and R, for any $I = 1, 2, \cdots R$, if R times the relevant inspection is completed, there is:

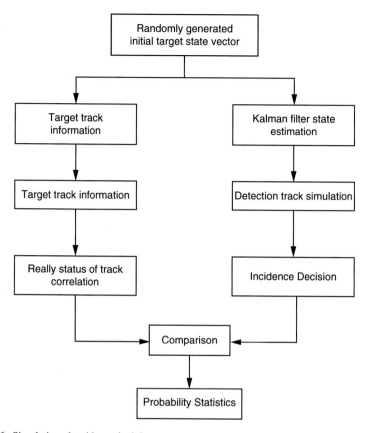

Fig. 5.6 Simulation algorithm principle

$$C_{ra}(R) > I \qquad\qquad (5.30)$$

The decision related to a fixed track is correct, as well as to stop between their relevant inspections in subsequent decisions. Then, it is possible to enter the post-Processing such as location data fusion.

If, after the R correlation test is completed, there is:

$$D_{ra}(R) \geq I \qquad\qquad (5.31)$$

It is determined that the two tracks are totally irrelevant, so we stop their inter-related tests in subsequent decisions. If the two cannot be fixed correctly to the related track, we need to go to the next correlation test cycle.

The simulation algorithm principle is shown in Fig. 5.6.

To analyze the performance of the algorithm, we performed Monte Carlo simulation with 50 iterations. Each time simulation comprised 12 steps, in the 100 batches of the target environment, considering the case where there are only two receiving stations, and each station can obtain the target position and speed information, so the target status information is six-dimensional. Receiving station 1 has coordinates $(X, Y, Z) = (90\,\text{km}, 60\,\text{km}, 0)$ and receiving station 2 has coordinates $(X, Y, Z) = (120\,\text{km}, 20\,\text{km}, 2\,\text{km})$. The time measurement accuracy of the receiving station is $\sigma_{\Delta t} = 10$ ns, bearing error $\sigma_\theta = 1^0$, and pitching angle $\sigma_\phi = 1^0$.

$$\hat{X}\left(l|l\right) = \left[\hat{x}\left(l\right), \hat{y}\left(l\right), \hat{z}\left(l\right), \dot{\hat{x}}\left(l\right), \dot{\hat{y}}\left(l\right), \dot{\hat{z}}\left(l\right)\right]$$

Target state equation is:

$$X(k) = \Phi X(k - 1) + GV(k - 1) \tag{5.32}$$

The measurement equation is:

$$Z(k) = H(k)X(k) + W(k) \tag{5.33}$$

The measurement vector is $Z = (x, y, z)^T$ and the measurement error covariance matrix is:

$$R(k) = \begin{bmatrix} \sigma_x^2(k) & 0 & 0 \\ 0 & \sigma_y^2(k) & 0 \\ 0 & 0 & \sigma_z^2(k) \end{bmatrix} \tag{5.34}$$

algorithm of fuzzy synthesis operation selection weighted average model, the weights vector of the corresponding initial values are: $a_1 = 0.6$ and $a_2 = 0.4$, formula $r_{kl} = \exp\left(-\tau_k\left(u_k^2/\sigma_k^2\right)\right)$ $k = 1, 2$ $l = 1, 2, 3$; the factor is determined based on k to decide on the similar membership of two tracks. In the simulation, we get $\sigma_k = 0.8$, $\tau_1 = 0.08$, $\tau_2 = 0.08$. The preset thresholds are $\varepsilon_1 = 0.1$, $I = 6$, and $R = 9$.

For an evaluation of the effectiveness of the association test, we consider three probabilities. The first category is the correct association probability E_c, with correct association of the probability from the same target with different nodes. The second category is the wrong association probability E_e, assigned to two tracks from different targets for the associated probability. The third category is the leakage association probability E_s, which represents that tracks i and j were originally from a target, but association test assigned track i with the uncorrect association, i leakage association, the probability of such an event is the leakage association probability. Then, obviously, there is $E_c + E_e + E_s = 1$. In track correlation analysis, the aim is to obtain E_c, E_e, and E_s. Analytical expression is difficult, so the simulation must select the relative frequency instead of the probability. Obtain $N = \max(n_{1l}, n_{2l})$, where n_{1l} and n_{2l} are, respectively, the associated track number of l time local node tracks 1 and 2. The number involved in the association test, N_c,

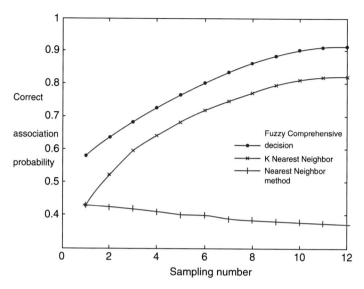

Fig. 5.7 Curve of the correct association probability

is the number of correct track associations, N_e is the number of wrong track associations, and N_s is the number of leakage track associations. Then:

$$E_c = {N_c}/{N_l}, E_e = {N_e}/{N_l}, E_s = {N_s}/{N_l} \qquad (5.35)$$

Apparently, for different l values, these results are different, but they increase with l. E_c, E_e, and E_s will be close to a stable value. From a random seed, select 100 of them. Figures 5.7 and 5.8 give comprehensive methods of judgment based on the fuzzy and the track association methods of nearest neighbor, where K is the wrong and correct association probability of the nearest neighbor track association method.

From Figs. 5.7 and 5.8, it can be seen that the track correlation algorithm of the fuzzy comprehensive decision simulation results is ideal. Over time, the probability curve stabilized and the correlation effects clearly improved.

The final decision is to meet the validation requirements of the track correlation. At the beginning, the correct association probability is low. Since for the simulation set the relevant threshold is unreasonable, with increasing sampling number, the correct association probability increases. With respect the statistical theory of the track correlation algorithm the probability of a fuzzy comprehensive decision for correct track association has a more significant increase.

When the fuzzy comprehensive decision is carried out on the track association processing, how to choose an appropriate rating, weighted distribution, membership, and related parameters such as threshold of association cause problems, because they have an effect on the performance and quality of the association.

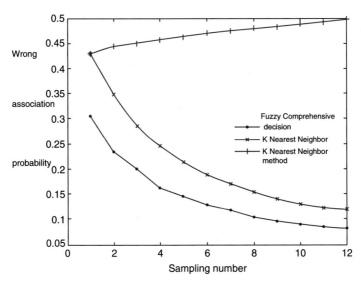

Fig. 5.8 Curve of the wrong association probability

Especially, the threshold of association is very important. If it is set high, the leakage association probability increases. On the contrary, the wrong association probability increases. For the determination of the association threshold value, we must perform a great many simulation experiments. Besides, if the target appears to be in maneuvering situations, an acceleration component must be introduced. If more track correlation discrimination information is introduced in the fuzzy rules, the accuracy of the track correlation discrimination is higher, but this will increase the associated fuzzy computational effort, reducing the discrimination speed of track correlation. The same is true for track correlation and size of the quality threshold value, which is also related to reliability problems association; if the value is large, the correlation reliability is higher, but the correlation time increases. Their value range is usually between 6 and 8.

5.3.3 Evidence Track Association Algorithm of Fuzzy Comprehensive Decision

The fuzzy based comprehensive decision track association algorithm has the advantage of good computing speed, etc. But in the design of the fuzzy model evaluation system, there is a lack of flexibility and versatility. Many parameters need to be set in advance, which yields lower intelligence, and the fuzzy integrated decision just belongs to a fuzzy set of membership, in the case of fully understanding the object under study. Such a conclusion is credible, but due to the uncertainty of the target motion state, it makes the object under study difficult to fully understand. Therefore, it is difficult to give the element that belongs to the fuzzy set membership.

In fuzzy track association, each index of the assessment value of the track association factors set is a fuzzy quantity. In the track association decision, this causes a large uncertainty. In the field of uncertainty, it is convenient, flexible, has a universal evidence theory, and many other useful features, so it has very good prospects, but in practical applications, its obvious shortcoming is that evidence is difficult to form. The concept of membership in the fuzzy comprehensive decision and the probability distribution in evidence theory are very similar in the physical sense, and comprehensive evaluation based on evidence theory and fuzzy comprehensive evaluation is based on that factor. Give indicators about each evaluation measure, then integrate the final conclusion. So, evidence theory with fuzzy comprehensive evaluation conducted in conjunction with track correlation judgment is a new way of thinking.

The evidence-track-association algorithm based on the fuzzy comprehensive decision absorbs and refers to the each index weighted coefficient distribution method of fuzzy comprehensive decision. It is similar to get fuzzy factor sets and fuzzy evaluation sets between the algorithm and the fuzzy comprehensive decision algorithm. The difference of the evidence-track-association algorithm based on the fuzzy comprehensive decision and the fuzzy decision is to determine the membership weighting coefficient and form evidence. The algorithm flow is shown in Fig. 5.9.

Fuzzy comprehensive decision for fuzzy factors involves a specified weighted coefficient. This approach is not flexible enough, and it cannot reflect the real-time track correlation degree of fuzzy factors, so you can consider the use of weighting to obtain evidence of membership, and then carry out fusion based on the Dempster–Shafe (D–S) evidence theory. But the traditional weighted criteria cannot be directly applied here, which is mainly caused by the uncertainty when there is a single factor, which makes a decision recognition framework of trust dominant. Even if the weight factor is very small, it may be due to weight and cannot make a correct decision. For example, recognition frameworks are, respectively:

$$\Theta = \{A, B, \theta\} \tag{5.36}$$

$$\begin{cases} m_1(A) = 0.7, m_1(B) = 0.3 \\ m_2(A) = 0.1, m_2(B) = 0.9 \end{cases} \tag{5.37}$$

θ means uncertain. The decision factors (weights) were 0.7 and 0.3, According to the weighted average fusion rule, do not use the D–S combination formula:

$$\begin{cases} m(A) = \alpha_1 m_1(A) + \alpha_2 m_2(A) = 0.52 \\ m(B) = \alpha_1 m_1(B) + \alpha_2 m_2(B) = 0.48 \end{cases} \tag{5.38}$$

In spite of the fusion results judgment A for narrowing down the final decision,

Fig. 5.9 The evidence
theory of the fuzzy
comprehensive decision
relevance principle diagram

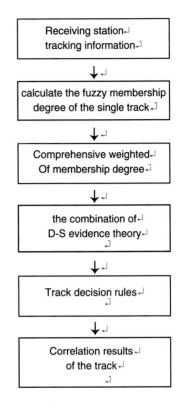

A and B have a gap of only 0.04, which easily produces system errors. They do not
meet the design criteria of the fuzzy comprehensive decision algorithm.

The determination method of the basic probability assignment function of multi-
feature information based on fuzzy comprehensive evaluation is available as:

$$m_{kl}(X_k) = a_k r_{kl}$$
$$m_{kl}(\theta) = 1 - m_{kl}(X_k)$$

(5.39)

The r_{kl} are membership functions of the fuzzy comprehensive track information
in decision making, a_k are the weighting coefficients, giving the contribution of
each feature of the evidence, and $\sum a_k = 1$. $m_{kl}(X_k)$ are the trust features to
distinguish the model belonging to this class. $m_{kl}(\theta)$ states that this part of the
trust cannot be determined.

The use of the D–S evidence combination rule yields:

$$m(H) = m_1 \bigoplus m_2 \bigoplus \cdots \bigoplus m_n$$

(5.40)

The reliability calculation after fusion supports the degree of fusion of the
reliability of the decision making, and in making decisions. At this time, the weight

is only related to the degree of uncertainty factors, which is $a_k + m_k(\theta) = 1$. Therefore, there is the need to modify the type, so the two weights were 0.7 and 0.3 for the initial evidence for the formation of the two decisions:

$$\begin{cases} \alpha_1 = 0.7, & m_1(A)' = \alpha_1 m_1(A) = 0.49, m_1(B)' = \alpha_1 m_1(B) = 0.21, m_1(\theta) = 1 - \alpha_1 = 0.3 \\ \alpha_2 = 0.3, & m_2(A)' = \alpha_2 m_2(A) = 0.03, m_2(B)' = \alpha_2 m_2(B) = 0.27, m_2(\theta) = 1 - \alpha_2 = 0.7 \end{cases}$$

$$\leq$$

$$(5.41)$$

The uses of the D–S evidence theory are combined, and the result is:

$$\begin{cases} m(A) = m_1 \oplus m_2(A) = 0.4103 \\ m(B) = m_1 \oplus m_2(B) = 0.3305 \\ m(\theta) = m_1 \oplus m_2(\theta) = 0.2438 \end{cases} \quad (5.42)$$

Relative to the direct weights, the final decision for the credibility of A must be high, but now there are still some problems; the trusts of B and A are worse than uncertainty. The trust of A and B is far less than the degree of uncertainty. The decision itself still has certain problems, so we consider the evidence to reconstruct, due to μ_k being the weights of fuzzy factors a_k. When multiple factors participate in the decision, the two following points should be considered:

1. On the right of the major factors, hoping that its probability distribution in evidence fusion plays a leading role, therefore, if there is a corresponding membership of association. According to the single factor judgment, the results should be related, then the weighting coefficient of the probability distribution is small relative to the weight of the factors of amplification. Conversely, when the corresponding association membership degree is small, then the weighting coefficient of the probability distribution of weight relative to the factors should shrink;
2. Compromise of small weight factors, with the same result when the membership is large, the probability distribution is reduced, as are the membership small with probability distribution amplification.

Based on the evidence considered in the above process, we are in front of r_k membership on the last α_{ki} to achieve the weighted coefficient of evidence, $\alpha_{ki} = f(\alpha_1, \alpha_2, \cdots \alpha_n,)$. Take:

$$\alpha_{ki} = \begin{cases} a_k & r_{ki} > \delta \\ a_{n+1-k} & r_{ki} \leq \delta \end{cases} \quad (5.43)$$

That is equivalent to the weighted fuzzy factors according to the degree of membership at the exchange, where δ is the single factor decision threshold. Its value can be determined according to the simulation. From here, $\delta = 0.5$.

For the definition of evidence matrix $M = (m_{kj})_{n \times 3}$, among them:

$$m_{k1} = m_{k1} \, (Associated) = a_{k1} \, r_{k1} \tag{5.44}$$

the k factor

$$m_{k2} = m_{k2} \, (notassociated) = \alpha_{k2} \, r_{k2} = \alpha_{k2} (1 - r_{k1}) \tag{5.45}$$

As for the first factor, the track is not associated with trust:

$$m_{k3} = m_{k3} \, (Notsure) = 1 - m_{k1} - m_{k2} \tag{5.46}$$

This trust degree track is a not sure relationship.

m_{k1} and m_{k2} are directly formed from the membership degree of evidence, and m_{k3} is determined according to the weight coefficient of the evidence. The explanation for m_{k3} starts with two extremes that infer the following:

When $m_{k1} + m_{k2} = 1$, $m_{k3} = 0$, namely, the contributing factor μ_k is only one statement, only consider m_{k1} and m_{k2} as two indicators that can make decisions;

When $m_{k1} + m_{k2} = 0$, $m_{k3} = 1$, when μ_k has no influence on decisions, according to the single factor of uncertainty in the decision that reached 1. It is said that, considering these factors will not bring any useful information for the decision.

In the case where α_k and α_{ki} are not consistent, it is discussed in the following:

In order to meet the offer from the start, according to the size of the weights to adjust the relationship of the probability distribution given by the various factors, and to make it convenient to discuss, set $n = 2$. At this point in the process of D–S evidence fusion there exist corresponding relations as follows:

$$
\begin{aligned}
m(\text{Associated}) = {} & m_1(\text{Associated}) m_2(\text{Associated}) \\
& + m_1(\text{Associated}) m_2(\text{Notsure}) \\
& + m_1(\text{Notsure}) m_2(\text{Associated})
\end{aligned} \tag{5.47}
$$

$$
\begin{aligned}
m(\text{Not associated}) = {} & m_1(\text{Not associated}) m_2(\text{Not associated}) \\
& + m_1(\text{Not associated}) m_2(\text{Not sure}) \\
& + m_1(\text{Not sure}) m_2(\text{Not associated})
\end{aligned} \tag{5.48}
$$

$$m(\text{Notsure}) = m_1(\text{Notsure}) m_2(\text{Notsure}) \tag{5.49}$$

We do not change the evidence for the weighted standards:

$$
\left\{
\begin{aligned}
& m_{k1}(\text{associated}) = \alpha_1 r_{11}, m_{k1}(\text{Not associated}) = \alpha_1 r_{12}, m_{k1}(\text{Not sure}) = 1 - \alpha_1 \\
& m_{k2}(\text{Associated}) = \alpha_2 r_{21}, m_{k2}(\text{Not associated}) = \alpha_2 r_{22}, m_{k2}(\text{Not sure}) = 1 - \alpha_2
\end{aligned}
\right.
\tag{5.50}
$$

According to a new weighted evidence:

$$\begin{cases} m_{k1}(\text{Associated}) = \alpha_{11}r_{11}, m_{k1}(\text{Not associated}) = \alpha_{12}r_{12}, m_{k1}(\text{Not sure}) = 1 - \alpha_{11} - \alpha_{12} \\ m_{k2}(\text{Associated}) = \alpha_{21}r_{21}, m_{k2}(\text{Not associated}) = \alpha_{22}r_{22}, m_{k2}(\text{Not sure}) = 1 - \alpha_{21} - \alpha_{22} \end{cases}$$
$$(5.51)$$

For the synthesis of evidence m, compare one type after another in the same way. The following are synthesized from formula (5.50):

$$m_{k1}(\text{Associated}) = \alpha_1\alpha_2 r_{11}r_{12} + \alpha_1(1 - \alpha_2)r_{11} + \alpha_2(1 - \alpha_1)r_{21} \qquad (5.52)$$

and the following are synthesized from formula (5.51):

$$m_{k2}(\text{Associated}) = \alpha_{11}\alpha_{21} r_{11}r_{12} + \alpha_{11}(1 - \alpha_2)r_{11} + \alpha_{21}(1 - \alpha_1)r_{21} \qquad (5.53)$$

Both minus for:

$$(\alpha_{11}\alpha_{21} - \alpha_1\alpha_2)r_{11}r_{21} + (\alpha_{11} - \alpha_1)(1 - \alpha_2)r_{11} + (\alpha_{21} - \alpha_2)(1 - \alpha_1)r_{21} \quad (5.54)$$

When $r_{11} > \delta$, we have $r_{21} < \delta$, then at this time:

$$\begin{cases} \alpha_{11}\alpha_{21} - \alpha_1\alpha_2 = \alpha_1\alpha_1 - \alpha_1\alpha_2 > 0 \\ \alpha_{11} - \alpha_1 = \alpha_1 - \alpha_1 = 0 \\ \alpha_{21} - \alpha_2 = \alpha_1 - \alpha_2 > 0 \end{cases} \qquad (5.55)$$

The result of formula (5.54) is greater than 0, so the correlation degree becomes larger.

When $r_{11} \leq \delta$, we have $r_{21} > \delta$, then at this time:

$$\begin{cases} \alpha_{11}\alpha_{21} - \alpha_1\alpha_2 = \alpha_2\alpha_2 - \alpha_1\alpha_2 < 0 \\ \alpha_{11} - \alpha_1 = \alpha_2 - \alpha_1 < 0 \\ \alpha_{21} - \alpha_2 = \alpha_2 - \alpha_2 = 0 \end{cases} \qquad (5.56)$$

The result of formula (5.54) is less than zero, so the correlation degree is reduced.

A similar reasoning can be proved, and a new weighted association began to put forward in the conditions, in line with its weights representing significance.

From formula (5.37), using the new weighted calculation of evidence synthesis problem presented above gives:

$$\begin{cases} m_1(A)' = \alpha_{11}m_1(A) = 0.49, m_1(B)' = \alpha_{12}m_1(B) = 0.09, m_1(\theta) = 0.42 \\ m_2(A)' = \alpha_{21}m_2(A) = 0.07, m_2(B)' = \alpha_{22}m_2(B) = 0.27, m_2(\theta) = 0.66 \end{cases} \qquad (5.57)$$

Using the D–S combination formula, the synthesis of evidence is:

$$\begin{cases} m(A) = m_1 \oplus m_2(A) = 0.5129 \\ m(B) = m_1 \oplus m_2(B) = 0.2167 \\ m(\theta) = m_1 \oplus m_2(\theta) = 0.2704 \end{cases} \tag{5.58}$$

Clearly:

$$m(A) - m(B) > m(\theta) \tag{5.59}$$

At this time, the decision algorithm is correct.

In the parameter setting section, there are 50 simulations by the Monte Carlo method, with each simulation have 12 step, giving a batch of 120 in the target environment. Take the threshold as $\delta = 0.5$. The D–S combination rule is calculated. The decision rules associated with quality, from the definition of quality and as the same as the Sect. 5.3.2 the method of evidence theory based on fuzzy synthetic decision and fuzzy comprehensive decision method is correct and the false association probability are shown in Figs. 5.10 and 5.11.

As can be seen from the simulation results, the error probability evidence track association algorithm of the fuzzy comprehensive decision has higher correct association probability but lower correct association probability than the original fuzzy comprehensive decision based on correlation. The low number of cases, especially in the simulation based on the correlation of the fuzzy comprehensive decision correct association probability, is associated with the error probability being are higher than that of the fuzzy comprehensive decision performance. It shows the effectiveness of new algorithm. This is because, in the determination of the evidence, the weight distributions were related to treatment based on the physical meaning, making each decision factor able to reasonably fulfil its role in

Fig. 5.10 Curve of the correct association probability

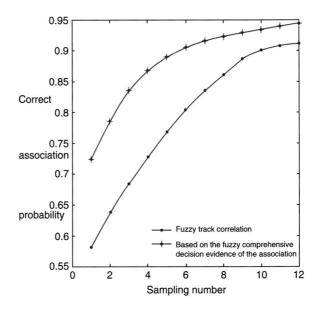

Fig. 5.11 Curve of the
false association probability

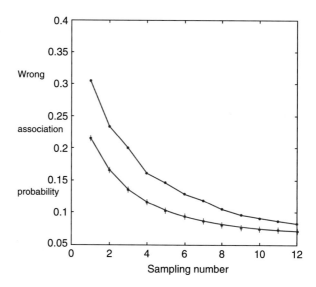

the judgment. Compared with the fuzzy comprehensive decision-making track correlation algorithm, the track correlation algorithm based on evidence theory can reach a satisfactory decision.

5.4 Track Fusion Network Radar Countermeasure System

Track association indicates that two tracks with high probability are from the same target, which is on track to track fusion has been on the track correlation in accordance with certain rules are combined to form a system. For a new track, the track state estimation and covariance calculation realizes the system track update. The center of a network radar countermeasure system performs the correlation, after the completion of the need for more concerns on the multiple combinations of the two local track information fusions. The result is system-unified tracking information. The common methods to track fusion include the simple fusion algorithm, cross-covariance combination algorithm, hierarchical fusion algorithm, covariance intersection algorithm, etc.

5.4.1 Simple Track Fusion and Cross-Covariance Combination Track Fusion

The simple track fusion (CC) algorithm was the first proposed track fusion algorithm. They are algorithms which assume that the local track state estimation error on the same target exhibit statistical independence. Its calculation is simple and the algorithm has been widely applied.

First, consider the reception track fusion by two stations, assuming the following target state model:

$$x_{k+1} = Fx_k + Gv_k \tag{5.60}$$

The receiving station $i \in [1, 2]$, at k time measurement $z_k^{(i)}$ obeys the linear model as follows:

$$z_k^{(i)} = H^{(i)}x_k + w_k^{(i)} \tag{5.61}$$

$v_k, w_k^{(i)}$ are statistically independent zero means with Gaussian white noise, and their variances are Q and $R^{(i)}$. Assuming $\hat{x}_{k|k}^{(i)}, P_{k|k}^{(i)}$ are i on the k receiving stations time target state minimum mean square error estimation and error covariance matrix. It fulfills the purpose of distributed fusion based on $\hat{x}_{k|k}^{(i)}, P_{k|k}^{(i)}$ optimization of the global route $\hat{x}_{k|k}, P_{k|k}$, but the main problems are the different receiving stations' track estimations of the correlation between the errors, which obey $P_{k|k}^{12} = P_{k|k}^{21} \neq 0$, and are unknowns. The cause of the error correlation includes: common process noise (considering the target mobility, Q is generally not zero). Estimate the prior common track $\left(\bar{x}_{k|k}, \bar{P}_{k|k} \right)$, caused by the special observation environment of different receiving stations' observation error correlation $\left(R^{(ij)} \neq 0 \right)$.

Simple track fusion error estimation assumes that different terminal tracks are statistically independent, and the fusion algorithm under the condition of two stations is:

$$\begin{aligned}
\hat{x}_{k|k} &= P_{k|k}^{(2)} \left(P_{k|k}^{(1)} + P_{k|k}^{(2)} \right)^{-1} \hat{x}_{k|k}^{(1)} + P_{k|k}^{(1)} \left(P_{k|k}^{(1)} + P_{k|k}^{(2)} \right)^{-1} \hat{x}_{k|k}^{(2)} \\
&= P_{k|k} \left(P_{k|k}^{(1)-1} \hat{x}_{k|k}^{(1)} + P_{k|k}^{(2)-1} \hat{x}_{k|k}^{(2)} \right)
\end{aligned} \tag{5.62}$$

$$P_{k|k} = P_{k|k}^{(1)} - P_{k|k}^{(1)-1} \left(P_{k|k}^{(1)} + P_{k|k}^{(2)} \right) P_{k|k}^{(1)} = \left(P_{k|k}^{(1)-1} + P_{k|k}^{(2)-1} \right)^{-1} \tag{5.63}$$

The algorithm exists only when the mutual covariance matrix $P_{k|k}^{(12)} = 0$ is optimal. If $P_{k|k}^{(12)} \neq 0$, it is just a sub-optimal approximate algorithm. As a result of the algorithm being unbiased, there is $P_{k|k} \left(P_{k|k}^{(1)-1} + P_{k|k}^{(2)-1} \right) = I$, where I is a unit matrix. If $P_{k|k}^{(12)} \neq 0$, the actual estimation error covariance for fusion is:

$$E\left[\left(x_k - \hat{x}_{k|k}\right)\left(x_k - \hat{x}_{k|k}\right)'\right] = P_{k|k} + P_{k|k}\left(P_{k|k}^{(1)-1}P_{k|k}^{(12)}P_{k|k}^{(2)-1} + P_{k|k}^{(2)-1}P_{k|k}^{(21)}P_{k|k}^{(1)-1}\right)P_{k|k}$$

$$(5.64)$$

While the CC algorithm completely ignores the error correlation, in the cross-covariance fusion combination track (BC) algorithm, it is considered in the effect of correlation on fusion estimation. The fusion algorithm for two stations is as follows:

$$\hat{x}_{k|k} = \hat{x}_{k|k}^{(1)}$$

$$+ \left(P_{k|k}^{(1)} - P_{k|k}^{(12)}\right)\left(P_{k|k}^{(1)} + P_{k|k}^{(2)} - P_{k|k}^{(12)} + P_{k|k}^{(21)}\right)^{-1}\left(\hat{x}_{k|k}^{(2)} - \hat{x}_{k|k}^{(1)}\right) \quad (5.65)$$

$$P_{k|k} = P_{k|k}^{(1)}$$

$$- \left(P_{k|k}^{(1)} - P_{k|k}^{(12)}\right)\left(P_{k|k}^{(1)} + P_{k|k}^{(2)} - P_{k|k}^{(12)} + P_{k|k}^{(21)}\right)^{-1}\left(P_{k|k}^{(1)} - P_{k|k}^{(21)}\right) \quad (5.66)$$

$$\hat{x}_{k|k} = F\hat{x}_{k-1|k-1}^{(i)} + K_k^{(i)}\left(z_k^{(i)} - H^{(i)}F\hat{x}_{k-1|k-1}^{(i)}\right) \quad (5.67)$$

In formula (5.67), $K_k^{(i)}$ is the filter gain. On the cross–covariance, $P_{k|k}^{(12)}$ has the following recursive relation:

$$P_{k|k}^{(12)} = E\left[\left(x_k - \hat{x}_{k|k}^{(1)}\right)\left(x_k - \hat{x}_{k|k}^{(2)}\right)'\right] = \left(I - K_k^{(1)}H^{(1)}\right)\left(FP_{k-1|k-1}^{(12)}F' + GQG'\right)$$

$$\times \left(I - K_k^{(2)}H^{(2)}\right)'K_k^{(1)}R^{(12)}K_k^{(2)},$$

$$(5.68)$$

In the above, $R^{(12)}$ is the two-terminal measurement error covariance. In order to calculate $P_{k|k}^{(12)}$, except for the state estimation information, so the receiving station also needs to pass the filter gain to the fusion center. Under steady-state conditions, $P_{k|k}^{(12)} = P_{k-1|k-1}^{(12)}$, formula (5.68) will evolve into a discrete Lyapunov equation. Because of the lack of a priori information, the BC algorithm is only in maximum likelihood (ML) rather than the minimum mean square error (MMSE) sense of the optimal estimate. Considering that the n receiving center station provides local tracking information, the definition is $\hat{X}_{k|k}^{loc} = \left[\hat{x}_{k|k}^{(1)}, \hat{x}_{k|k}^{(2)}, \cdots \hat{x}_{k|k}^{(N)}\right]'$, where $E = [I, I, \cdots I]'$ is a matrix of dimensions $Nn \times n$ and I is a unit matrix of dimensions $n \times n$.

$$P = \begin{matrix} P^{(11)}_{k|k} & P^{(12)}_{k|k} & \cdots & P^{(1N)}_{k|k} \\ P^{(21)}_{k|k} & P^{(22)}_{k|k} & \cdots & P^{(2N)}_{k|k} \\ \cdots & \cdots & \ddots & \cdots \\ P^{(N1)}_{k|k} & P^{(N2)}_{k|k} & \cdots & P^{(NN)}_{k|k} \end{matrix}$$

$$\hat{x}_{k|k} = \left(E'P^{-1}E \right)^{-1} E'P^{-1}\hat{X}^{loc}_{k|k} \tag{5.69}$$

$$P_{k|k} = \left(E'P^{-1}E \right)^{-1} \tag{5.70}$$

5.4.2 Covariance Intersection Fusion

The covariance intersection algorithm (CI) is based on the thought process of: If each $P^{(ij)}_{k|k}$ between the local state estimation covariance matrix is known, the fusion estimation of the covariance of an oval will be located in the local estimation covariance elliptical cross area, as shown in Fig. 5.12. In this figure, the inner ring of the solid line are the corresponding different $P^{(ij)}_{k|k}$ when the fusion estimation covariance matrix is elliptic, while the outer ring of the two solid lines of an oval are on behalf of the local estimate covariance matrix. Thus, though we do not know $P^{(ij)}_{k|k}$, the containing cross area must be consistent with the target state estimation, and it is better surrounded by the closer fusion performance. The dotted oval in the figure represents better satisfaction of the consistency estimation fusion estimation error covariance matrix. The CI algorithm is able to estimate from the consistency all the best choices as a fusion estimation, as shown by the dashed ellipse in Fig. 5.13.

Target states are assumed to be of a Gaussian distribution, then the two receiving stations case in the CI algorithm is as follows:

$$P^{-1}_{k|k}\hat{x}_{k|k} = \omega P^{(1)-1}_{k|k}\hat{x}^{(1)}_{k|k} + (1-\omega)P^{(2)-1}_{k|k}\hat{x}^{(2)}_{k|k} \tag{5.71}$$

$$P^{-1}_{k|k} = \omega P^{(1)-1}_{k|k} + (1-\omega)P^{(2)-1}_{k|k} \tag{5.72}$$

where ω is the weight of the local estimation. By minimizing the determinant of $P_{k|k}$ or if the trace is obtained, it can be prove that, for any $\omega \in [0, 1]$ and $P^{(12)}_{k|k}$, the fusion estimations are consistent. Obviously, the CI algorithm with the CC form is

Fig. 5.12 Schematic
diagram of 1 CI algorithm

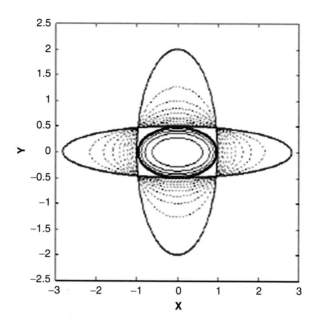

Fig. 5.13 Schematic
diagram of 1 CI algorithm

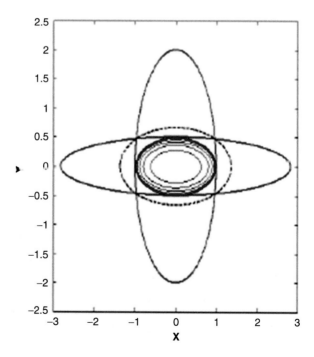

very similar, but with $P^{(1)}_{k|k}$ instead of $\omega^{-1}P^{(1)}_{k|k}$ and $P^{(2)}_{k|k}$ instead of $(1-\omega)^{-1}P^{(2)}_{k|k}$.

Generally, it is thought that the CC algorithm for fusion performance is too optimistic, given that $0 \leq \omega \leq 1$, the CI algorithm's fusion estimation covariance with the CC algorithm is relatively more conservative, such as when $\omega = 0.5$, like $\hat{x}^{CI} = \hat{x}^{CC}$ and $P^{CI} = 2P^{CC}$. When $P^{(1)} = P^{(2)}$, like $P^{CI} = P^{(1)} = P^{(2)}$, the CI algorithm did not reduce the uncertainty of the fusion estimation. The CI algorithm in multi-station $(N \geq 2)$ form yields:

$$P^{-1}_{k|k}\hat{x}_{k|k} = \omega_1 P^{(1)-1}_{k|k}\hat{x}^{(1)}_{k|k} + \ldots + \omega_N P^{(N)-1}_{k|k}\hat{x}^{(N)}_{k|k} \tag{5.73}$$

$$P^{-1}_{k|k} = \omega_1 P^{(1)-1}_{k|k} + \ldots + \omega_N P^{(N)-1}_{k|k} \tag{5.74}$$

In the above, $\sum \omega_i = 1$ are the weights of ω_i, which make the determinant of the fusion covariance matrix $P_{k|k}$ take the smallest value.

Track correlation and fusion are important parts of the system data processing center staions get local target information obtained from various terminals. They enable the association and integration of target tracking information.

Chapter 6
Four Countermeasure Capacity Analysis of Network Radar Countermeasure Systems

A network radar countermeasure system is an integration of radar and radar countermeasures of an integrated electronic warfare system. In addition to realizing the target reconnaissance detection, electronic jamming, and identification function, the system also has good anti-jamming, anti-counter radiation weapon damage, anti-stealth, and anti-low-altitude defense penetration ability.

6.1 Network Radar Countermeasure System Anti-jamming Performance Analysis

Active jamming is the use of special electronic jammers and equipment, which takes the initiative to transmit or transpond some form of jamming signal, to suppress, disrupt, or deceive the enemy radar. It is a kind of electronic jamming that makes enemy radar unable to function properly, which is also known as active jamming. Passive jamming is the use of chaff, corner reflector, and other special equipment to reflect, attenuate, or absorb radar radiation electromagnetic waves, disrupt the radar electromagnetic wave propagation, change the radar scattering characteristics of the target, or form false radar targets and interfere with the barrier. It is a form of passive jamming to cover the real target.

For the network radar countermeasure system, it is easy to eliminate passive jamming through data fusion technology. From the perspective of a target reconnaissance detection system, the influence of active jamming on the system is the focused here.

© National Defense Industry Press, Beijing and Springer-Verlag Berlin Heidelberg 2016   273
Q. Jiang, *Network Radar Countermeasure Systems*,
DOI 10.1007/978-3-662-48471-5_6

6.1.1   Anti-active Blanket Jamming

Effective blanket jamming of radar must meet the needs of four basic conditions. First, it must intercept the radar signal; second, the jamming antenna must point in the direction of the radar receiver antenna; third, the jamming signal frequency must cover the radar working frequency; fourth, the jamming power must reach the power that implements the required effective jamming.

This chapter will take the mono-static radar and a radar network as the reference, and analyze the capacity of the network against a radar reconnaissance system to resist active blanket jamming.

6.1.1.1   The Performance of Anti-active Oppressive Jamming Resistance with Passive Mode

When the system works in passive mode, due to there being no electromagnetic radiation signals, the radar countermeasure receiver of the other party is unable to carry on with the reconnaissance intercepted, so the prerequisite to implement effective jamming is destroyed. That, in comparison with the mono-static radar networking system, has stronger system concealment and resistance to intercept and anti-jamming performance. The netted radar stealth performance enhancement system method is to use the network radars switched on in turn to reduce the single machine work time, thereby reducing the exposure probability of the radar. However, in order not to affect radar target detection and tracking, at least one radar boot is needed. Therefore, for the passive reconnaissance model in concealment, and resistance to intercept and anti-jamming have all the advantages of the passive radar. It is better than the mono-static radar networking in terms of hidden features, and its anti-jamming ability is greatly increased.

6.1.1.2   The Performance of Anti-active Oppressive Jamming Resistance with Active Working Mode

The active work mode is another important mode in the system, which is different from passive radar. However, when the system is working in active mode, due to its different deployment, the transmitter, receiver, and the other party's radar receiver can intercept radar signals in the direction of the transmitter, so its jamming antenna pointing in that direction can be a transmitter, unable to direct the implementation of jamming in the receiver. It is a network radar reconnaissance system in principle, therefore, destroying the implementation of effective jamming of the second condition, so it is unable to realize jamming in the direction of the radar receiver.

In the direction of jamming at the transmitter, in order to reach the target of oppressive jamming, there is the need to use the noise jammer signal to cover or submerge the useful radar echo signal, thus, placing high demands on the jamming power. In order to form effective jamming, the power must be increased by dozens of decibels.

In quantitative terms, below we will show a network against radar reconnaissance system in the active mode for suppressing jamming. The effective detection range of radar is used for reference, to compare several radar systems under the same conditions when the noise disturbance detection performance is degraded.

First of all, we analyze the various factors that affect the jamming power received by a radar. The jammer receiver and radar jamming power are related as follows:

$$J_0 = \frac{P_j G_j G_R F_j^2 \lambda^2}{(4\pi)^2 B_j R_j^2 L_j} \tag{6.1}$$

In the formula, P_j is the jammer transmission power; G_j is the jammer antenna gain; $F_j = F_j' f_{jR} f_{Rj}$ is the propagation factor for the direction figure of the jammer; f_{jR} for the radar receiver base antenna pattern as a benchmark of jamming factors; f_{Rj} is the antenna pattern factor based on the jammer radar receiver; F_j' considers the spread of the multi-path, diffraction, and reflection factor; G_R is the radar receiving antenna power gain; B_j is the jammer bandwidth; R_j is the distance between the jammer and the radar receiver; and L_j is the jammer system loss.

For repressive noise jamming, the jammer bandwidth is bigger than the radar bandwidth. In the general case, $F_j' = 1$, so, in order to simplify the discussion, it is not considered to influence the other parameters, so we make it a constant one of 1 directly. So, the largest type structures with radar configuration in the relationship are f_{Rj} and f_{jR}.

According to the previous radar counter reconnaissance system detection performance analysis of a network, the active work mode of the calculation of the detection ranges of the system involve decomposition of a bi-static radar unit, so the impact analysis is to suppress jamming on the system detection performance. First, we discuss a bi-static radar with a mono-static radar detection range in the noise suppression disturbance problem. According to the relative operational situation of the jammer and the radar, there are four main geometric relationships, as shown in Fig. 6.1.

In order to enhance the jamming effect, the radar jammer is always the main beam of jamming, so the radar antenna is usually pointing the main beam to the target. So, for the mono-static radar against jamming, there will appear in Fig. 6.1a, b two cases: (a) the graph shows self-defense jamming, where jamming occurs on the radar antenna's main lobe, also known as main lobe jamming; (b) to support jamming, the jammer and the target are not located together, the jammer antenna's main lobe is on the radar antenna, known as side lobe jamming. For a bi-static radar and the radar countermeasures reconnaissance system, the enemy spy plane cannot judge the nature of the radar transmitter, and it is also unable to detect the location of the radar receiver. Therefore, it can only aim at the radar jamming source and,

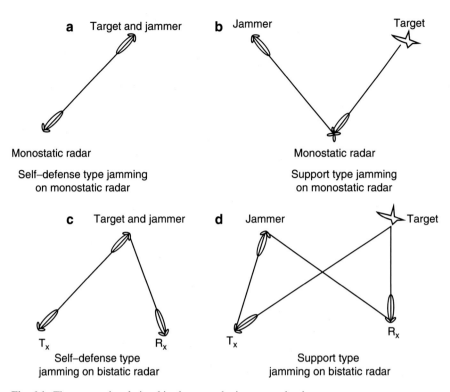

Fig. 6.1 The geometric relationships between the jammer and radar

therefore, there will be two cases as shown in Fig. 6.1c, d: (c) for self-defense jamming, side lobe alignment of the jammer and now the radar antenna's main lobe jamming; (d) to support jamming, the side lobe on the bi-static radar jammer or network against radar reconnaissance system receiving antenna undergo side lobe jamming. However, there are special circumstances. The target and jammer are situated on the baseline extension in the bi-static radar, and the main lobe of the bi-static radar can receive main lobe jamming.

Under the same jamming parameters, different war situations have different jamming effects. For the mono-static radar, the jamming circumstances are: (1) the jammer antenna main lobe jamming the radar antenna main lobe when referred to as the main lobe jamming, have $f_{jR} = f_{Rj} = 1$; (2) the main lobe of the jammer to the jamming radar side lobe, side lobe jamming for short, then the relationship is: $f_{jR} = 1, f_{Rj} < 1$. For the bi-static radar, the jamming circumstances are: (1) jammer antenna side lobe jamming the radar receiver antenna main lobe, referred to as double-base main lobe jamming, at this time meets the following: $f_{jR} < 1, f_{Rj} = 1$; (2) the jammer uses the side lobe to jam the bi-static radar receiver side lobe, double-base side lobe jamming for short, then: $f_{jR} < 1, f_{Rj} < 1$. After determining the relationship between the jammer and the radar, we can, according to formula (6.1), calculate the jamming power spectral density, and the performance of radar detection for a set jamming power can also be calculated. When there are multiple jammers jamming one radar at the same time, the total jamming power spectral density is as follows:

$$\sum J_0 = \sum_{i=1}^{n} (J_0)_i \tag{6.2}$$

Next, we will research the influence of noise effects on the performance of radar detection. Under the effect of noise, the input noise of the radar receiver is from N_0 up to $N_0 + \sum J_0$. If it shows as a noise temperature, then:

$$T'_s = T_s + \sum J_0/K \tag{6.3}$$

In the formula, T'_s is the noise temperature of the receiver under the circumstances of jamming; T_s is the thermal noise temperature of the system without jamming; K is the Boltzmann constant. T_S' instead of the radar equation of T_S is for a mono-static radar, which is:

$$
(R_M)_j = \left[\frac{P_T G_T G_R \lambda^2 \sigma F_j^2 F_R^2}{(4\pi)^3 \left(KR_s + \sum J_0 \right) B_n (S/N)_{min} L} \right]^{1/4}
$$

$$
= R_M \left[\frac{KT_s}{\left(KT_s + \sum J_0 \right)} \right]^{1/4} \tag{6.4}
$$

For a bi-static radar, it is:

$$
(R_T R_R)_j = \left[\frac{P_T G_T G_R \lambda^2 \sigma_B F_j^2 F_R^2}{(4\pi)^3 \left(KT_s + \sum J_0 \right) B_n (S/N)_{min} L} \right]^{1/2}
$$

$$
= k_B \left[\frac{KT_s}{\left(KT_s + \sum J_0 \right)} \right]^{1/2} = k_{Bj} \tag{6.5}
$$

In the formula, $(R_M)_j$ is the largest mono-static radar range under the conditions of the jamming, $(R_T R_R)_j$ is the bi-static radar range under the conditions of the jamming, and $k_B = R_T R_R$ is the bi-static radar range without the conditions of the jamming.

For bi-static radar transmitter and receiver technology and the equivalent mono-static radar, the transmitter and receiver parameters are the same. Comparing before and after the biggest mono-static radar jamming effect distance and bi-static radar range product, we can see the characteristics of the bi-static radar jamming suppression of resistance; thus, the advantages of resistance to suppress jamming can be seen from the network against radar reconnaissance system.

Assume that the parameters of the jammer are $P_j = 500\,\text{W}$, $G_j = 16\text{dB}$, $B_j = 500$ MHz, and $L_j = 3\text{dB}$; The technical parameters for the radar are: $\lambda = 0.1\text{m}$, $G_R = 33\text{dB}$, $kT_s = 6 \times 10^{-21}\,\text{W/Hz}$, $R_M = 100\,\text{km}$, $kB = 10000\text{km}^2$, $R_j = 100\,\text{km}$, and the bi-static radar baseline $L = 100\,\text{km}$. Then, according to the relationship between several kinds of geometric layouts as shown in Fig. 6.1, the corresponding results can be obtained:

1. Self-defense jamming effects on a mono-static radar, which interfere with the main lobe of the radar main lobe jamming. If $f_{jR} = f_{Rj} = 1$, then:

$$J_0 = \frac{P_j G_j G_R \lambda^2}{(4\pi)^2 B_j R_j^2 L_j} \tag{6.6}$$

Through calculation, under the setting parameters, $J_0 = 2.5 \times 10^{-16} W/Hz$. So, disturbances after the maximum range of the radar are as follows:

$$[R_M]_j = R_M \left[\frac{KT_s}{KT_s + J_0} \right]^{1/4} = \left[\frac{6 \times 10^{-21}}{6 \times 10^{-21} + 2.5 \times 10^{-16}} \right]^{1/4} = 0.07 R_M = 7km \tag{6.7}$$

The range of a mono-static radar after jamming decreased to 7 %, the maximum range is only 7 km, and the scope is only 153 km^2.

2. When the jamming of a mono-static radar is supported (that is, interference with the main lobe to side lobe jamming radar receiver), it is assumed that the side lobe level radar for −3 dB is now $f_{jR} = 1$, $f_{Rj}^2 = -36dB$, then:

$$J_0 = \frac{P_j G_j G_R \lambda^2 f_{Rj}^2}{(4\pi)^2 B_j R_j^2 L_j} \tag{6.8}$$

Plugging in the corresponding parameters, it is calculated that $J_0 = 6.3 \times 10^{-20} W/Hz$. With disturbance, the maximum range of the radar is:

$$[R_M]_j = R_M \left[\frac{KT_s}{KT_s + J_0} \right]^{1/4} = \left[\frac{6 \times 10^{-21}}{6 \times 10^{-21} + 6.3 \times 10^{-20}} \right]^{1/4} = 0.56 R_M$$
$$= 56km \tag{6.9}$$

3. Self-defense jamming of bi-static radar jamming, namely, the jammer side lobe jamming the radar main lobe, assumes that the side lobe level of the jammer is −6 dB, it is $f_{Rj} = 1$, $f_{jR}^2 = -22dB$, so the following formula was established:

$$J_0 = \frac{P_j G_j G_R \lambda^2 f_{jR}^2}{(4\pi)^2 B_j R_j^2 L_j} \tag{6.10}$$

Putting the corresponding parameters into the above formula, it is calculated as $J_0 = 1.59 \times 10^{-18} W/Hz$. At this time, the bi-static radar range product is as follows:

$$(R_T R_R)_j = k_B \left[\frac{KT_s}{\left(KT_s + \sum J_0 \right)} \right]^{1/2} = k_B \left[\frac{6 \times 10^{-21}}{6 \times 10^{-21} + 1.59 \times 10^{-18}} \right]^{1/2} = 612.4 km^2 \tag{6.11}$$

4. Support jamming of bi-static radar jamming, namely, the jammer side lobe jamming the side lobe of the radar receiver, it is $f_{Rj}^2 = -36dB$, $f_{jR}^2 = -22dB$, then:

$$J_0 = \frac{P_j G_j G_R \lambda^2 f_{Rj}^2 f_{jR}^2}{(4\pi)^2 B_j R_j^2 L_j} \tag{6.12}$$

Substitution of the calculated parameters results in $J_0 \approx 4 \times 10^{-22} W/Hz$. At this time, the product range of the bi-static radar is as follows:

$$(R_T R_R)_j = k_B \left[\frac{KT_s}{(KT_s + J_0)} \right]^{1/2} = 0.97 k_B = 9700 \, km^2 \tag{6.13}$$

As can be seen, in this case, the scope of the bi-static radar basically remains unchanged; that is to say, the bi-static radar has a strong anti-jamming ability.

From the above results, it can be seen that a bi-static radar, in active noise suppressing jamming, has very strong ability, and the network against radar reconnaissance system active detection is the basis of a bi-static radar; therefore, has stronger anti-jamming ability. In addition, in the case of jamming, the radar should not wait for the burn-through distance and find its target. It should find the target from longer distances, and guide the weapons system to destroy it. The network against radar reconnaissance system can use jammer radiation sources to position targets and jammers, which greatly increases the radar detection range and saves a lot of time for guiding weapons attacks.

6.1.2   Resistance to Active Deception Jamming

Active deception jamming is another important type of active jamming. The expected effect of deceptive jamming is to produce false targets, mixing false with genuine targets, and cheat or confuse radar. Its basic principle is the use of reconnaissance equipment to detect radar signals, through digital radio frequency memory (DRFM) technology or other means to replicate the same or similar jamming signals as the radar signal, after appropriate delay, modulation, and amplifying processes, then send them out, making the radar tracking find or display false targets, so as to deceive. Jamming the warning radar with repeater jammer false targets is not easy to distinguish because the false targets and the target signals are mixed together, providing wrong information to the enemy surveillance radar, confusing the enemy to a great extent and its troops dispersed. Deceptive jamming come into its element in the distance traction, angle traction, and speed traction to the tracking radar, increasing the tracking error and even causing the tracking target to be lost completely.

Effectively deceptive jamming to the radar must meet five basic conditions: first, it must intercept the radar signal; second, the jamming antenna pointing direction must be on the radar receiver antenna; third, the jamming signal frequency must be aimed at the working frequency of the radar signal; fourth, the jamming signal power must be greater than the target echo, with the general requirement that the jamming signal ratio is more than 10 dB; fifth, the jamming signal waveform must be similar to the radar signal, the so-called similar jamming signal should have essentially the same form as the radar signals, and at the same time, have a certain amount of false information. The feature parameters of the radar part are the same, making it difficult to distinguish some characteristic parameters, such as the radar extraction error information.

In reference to the mono-static radar and mono-static radar networking, assume that the implementation of active deception jamming against several trend diagrams is shown in Fig. 6.1, comparing network radar countermeasure system active suppression of resistance against jamming.

6.1.2.1   The Performance of Anti-active Deception Jamming with Passive Working Mode

Deception jamming is through effective radar radiation signal capture, storage, copying, and forwarding to realize the radar deception jamming. The precondition is to intercept and analyze the characteristic parameters of the radar signal, and then, according to the purpose of jamming, modulate one or several parameters of signal and form false information to send out, to leverage radar deception jamming.

It can be seen that the premise of deception is that the radar must radiate a signal. In order to accomplish the detection and tracking tasks, a mono-static radar must make the radar work and constantly radiate electromagnetic signals across the airwaves, but this provides an opportunity to use radar jamming reconnaissance equipment. The radar parameters under the condition of modern technology such as extraction signal copy is no longer a problem, so a radar deception jamming probability increases greatly. Although with the development of modern radar there also appeared some new radar systems, inhibiting the threat of radar jamming to a certain extent, but with the development and maturing of DRFM technology, radar deception jamming threats are further expanding again. So, radar networks become an important development direction to improve anti-jamming ability. Netted radar increased the difficulty in radar jamming reconnaissance, the radar jamming signal formation caused some obstacles, and the use of information fusion technology in netted radar improved, to a certain extent, the ability of deceiving jamming signal recognition. But it is not to say that the netted radar can't provide jamming. For mono-static radar networking, its covert way is to enhance the system by changing the radar rotating time to reduce the working time of a single radar, reducing the exposure probability of the system. But reducing the probability of exposure, after all, still has a risk of exposure, such as in the previous example, where a network of three same radars operate, then a single radar unit's exposure probability induces a maximum reduction of 1/3. In war, if the enemy increases their radar jamming equipment, jamming the mono-static radar of a netted radar still has a jamming effect.

When the system is working in the passive detection mode, it will not radiate any form of electromagnetic radiation energy, so the premise condition of disturbances is destroyed, and deception jamming cannot be imposed. In addition, the radar jamming is often based on the radar threat level to choose the object, generally, in combat, of tracking the radar threat level to be higher than the detection radar, and for tracking radar deception jamming, the probability is higher. In this mode, the system of target detection, location, tracking, and other working modes in the transmission of the signal form is no difference, so it generally cannot attract adverse attention, further reducing the probability of jamming.

6.1.2.2 The Performance of Anti-active Deception Jamming with Active Working Mode

The characteristics of the system to receive and dispatch separation and multi-station configuration in the inhibition of deception jamming also have great advantages. First, one of the essential requirements of deceptive jamming effects is to interfere with the antenna in the direction of the radar receiver. The sending and receiving characteristics of a network radar countermeasure system are different in deployment, and the jamming antenna cannot aim in the receiver's direction. Second, due to the sending and receiving of different deployments, it is difficult to realize the angle, distance, and velocity deception. Third, it can produce some kind of deception, because deceiving the jamming power is also many times greater than for a mono-static radar.

From the standpoint of power, when jamming on a single radar of the mono-static netted radar, the situation is generally the jammer main lobe jamming on the radar receiver antenna's main lobe, so the jamming has obvious superiority in power. The transmitter station and the receiver terminal of the network radar countermeasure system are separated, so the jamming situation is generally in self-defense jamming for the side lobe of the jamming station jamming the main lobe of the radar receiver, while supporting jamming for the jamming station side lobe jamming the side lobe of the radar receiver. Therefore, in radar deception jamming against the reconnaissance system in the network, the jamming power loss is very large, inhibiting the jamming effect to a certain extent. Here is a simple example to illustrate the problem.

The jamming power received by the radar receiver is:

$$J = \frac{P_j G_j G_R F_j^2 \lambda^2}{(4\pi)^2 R_j^2 L_j} \tag{6.14}$$

The received radar echo signal power is:

$$S = \frac{P_T G_T G_R \sigma \lambda^2 F_T^2 F_R^2}{(4\pi)^3 R_T^2 R_R^2 L_T L_R} \tag{6.15}$$

So, the radar receiver the signal to jamming ratio is as follows:

$$\frac{S}{J} = \frac{P_T G_T}{P_j G_j} \cdot \frac{F_T^2 F_R^2}{F_j^2} \cdot \frac{L_j}{L_T L_B} \cdot \frac{\sigma}{4\pi} \cdot \frac{R_j^2}{R_T^2 R_R^2} \tag{6.16}$$

In the above, P_T is the radar transmitted power, G_T is the radar antenna main lobe power gain, G_R is the radar receiving antenna main lobe gain, σ is the scattering cross-section of the radar target, and the other parameters are the same as those from the definition at the beginning. In computing the jamming ratio of the network radar countermeasure system, the network against radar reconnaissance system terminal receives the echo signal power and signal power, so the calculation method of the bi-static radar can be used. Then, according to the different enemy situations, the signal to jamming ratio can be obtained. In jamming the mono-static radar, the following conditions are met $G_T = G_R$, $\sigma = \sigma_M$, $L_T = L_R$, $F_T = F_R$, and L_j contains the bandwidth loss of the deception jammer. In general, the deception jammer is the target carrying the self-defense jammer and meets $F_j = F_R = F_T = 1$ and $f_{Rj} = f_{jR} = 1$, $L_T = L_R = L$, $\sigma = \sigma_M$, so the signal to jamming ratio of the mono-static radar is:

$$\left(\frac{S}{J}\right)_M = \frac{P_T G_T}{P_j G_j} \cdot \frac{L_j}{L^2} \cdot \frac{\sigma_M}{4\pi R_T^2} \tag{6.17}$$

For the bi-static radar, we still have $R_j = R_R$, $F'_T = F'_j = F'_B$, but $f_{Rj} = 1, f_{jR} < 1$, $\sigma = \sigma_B$. At this time, the radar receiver signal to jamming ratio is as follows:

$$\left(\frac{S}{J}\right)_B = \frac{P_T G_T}{P_j G_j} \cdot \frac{1}{f^2_{jR}} \cdot \frac{L_j}{L_T L_R} \cdot \frac{\sigma_B}{4\pi R_T^2} \tag{6.18}$$

By comparing the above two types of signal to jamming ratio, it can be seen that, for the bi-static radar with the same technical level compared to the mono-static radar, the equivalent mono-static radar has a certain advantage in resistance to deception jamming. And the network radar countermeasure system at this time can be equivalent to more than one common form of bi-static radar network. It will also help each other to share information and data, further eliminating false information, so its ability to resist deception jamming is stronger. Of course, making the mono-static radar network formation a netted radar also improves the ability of the radar deception jamming resistance to a certain extent, and can be conducted in the center of the data fusion, to remove part of the false target. But by comparison, the basic unit in deception jamming resistance performance has been significantly weaker than the network radar countermeasure system, as the data fusion and network protocols are not as flexible as a network against radar reconnaissance system; this further limits its ability to resist deception jamming.

6.2   The Performance Analysis of Network Radar Countermeasure System on Anti-stealth

Stealth technology was the most important technology during the 1980s military aviation technology breakthrough, and is also an important milestone in the history of the development of aviation technology. During the Gulf War, the United States used 42 F-117A stealth fighters to break through the Iraqi national military command center and the air force command structure around the intensive camouflage net, because the F-117A stealth fighter is the only aircraft that can have free flight in fortified overhead airspace over Baghdad. In 1999 during the Kosovo War, NATO bombed Yugoslavia and used a variety of stealth weapons and missiles. In a few days of air strikes, in addition to using the F-117A stealth fighter and AGM-129 stealth cruise missiles, NATO also used B-2 stealth bombers for the first time. The cross-section (RCS) of a B-2 stealth bomber's radar is only 0.1 m^2, and those of an F-117A stealth fighter and AGM-129 stealth cruise missile RCS are smaller. The reduced target RCS greatly shortens the radar detection range for these targets, thus greatly enhancing the stealth aircraft air penetration ability. The emergence of stealth technology changed the face of modern air force and air combat modes, putting forward serious challenges for modern radar.

Radar target stealth technology mainly includes shape design, absorbing materials, active noise cancellation, etc. The different kinds of stealth techniques are not identical in stealth effects, which are comprehensive in using these techniques, and can make the aircraft's radar reflection sectional area reduce to levels of about 20 dB or more. Radar anti-stealth faces three problems: (1) the appearance of unique designs and load impedance technology to cause anti-stealth difficulties; (2) radiation-absorbent material (RAM) is used in radar-absorbing structures to increase the difficulty of anti-stealth; (3) the anti-stealth radar should be considered in the "three anti" requirements. Visible stealth technology has become a severe challenge for radar to play out its effectiveness.

6.2.1   Passive Work Mode for Anti-stealth

At present, the main methods used in radar target stealth aim at enemy active radiation. It uses all kinds of technology to reduce the target direction of radar backscatter echo, thereby reducing enemy radar detection probability, such as shape design, absorbing materials, active offset, etc. But in order to implement their own detection or communication, the stealth target itself must radiate electromagnetic signals. The network against radar reconnaissance system works in passive mode. The system itself does not radiate electromagnetic signals, but it can intercept and analyze the signals to obtain the location of the object and its attribute information through the target's radiation of electromagnetic waves.

According to intercepted stealth target emitter signals, the measurement and analysis of the main features of the radiation source parameters, real-time identification, and determining target attributes, it can provide detailed technical parameters about the targets and individual characteristics, and point trace related, path tracking technology of the real-time target track formation and the same air defense early warning, to complete the conventional radar guidance function. Compared with traditional active detection, the passive mode also has the role of far distance, strong concealment, not easy to be found, and other advantages. The passive mode can be used at farther distances from the enemy target; thus, it has good anti-stealth ability.

6.2.2 Selection of Proper Frequency Increasing Anti-stealth Capability

Stealth technology adopts several main measures that are closely related to the wavelength of the incident electromagnetic wave. When the radar wavelength is close to the size of any part of the target, electromagnetic resonance can be produced in this part causing strong reflection. In addition, because of aerodynamics, absorbing material, and frequency response limitations, the aircraft is stealthy within the 1–20 GHz frequency range. For frequency of defects, and anti-stealth radar materials, the new system of harmonic radar mainly includes meter wave radar, over-the-horizon radar, shock wave radar, and the millimeter wave radar, and acceptable metal targets twice to three times the ability of harmonic signals etc., but it is limited by a variety of things in practice. For the meter wave radar, its signal processing ability, beam width, low measuring precision and resolution, and signal bandwidth is generally narrower, and it has poor anti-jamming ability. Over-the-horizon radar stealth aircraft cannot give precise locations and have measuring errors of tens of kilometers distance. The only surety is the existence of stealth aircraft roughly the area, and its detection performance changes randomly with the ionosphere height and parameters, which need to be equipped with the ability to carry ionosphere real-time monitoring equipment and powerful computer computing ability. The Achilles' heel of the shock pulse radar is low energy radiation, and the detection range is limited. Millimeter wave radar has the problem of the atmospheric attenuation being severe, causing serious shortening of the detection range.

The basic starting point of the mono-static radar networking in the frequency domain anti-stealth is the use of stealth target of VHF/UHF frequency band wave radar, and the HF band of the over-the-horizon radar stealth effect is poor. So there is a suggest to use the VHF/UHF frequency band and HF band, multiband radar, and radar network to improve the ability of anti-stealth. For netted radar, besides multiband measures, the multi-station long-distance deployment also weakens the target stealth effect, so the anti-stealth effect is much better than that of the mono-static radar.

Network against radar reconnaissance system in respect of stealth target detection can have the advantage of a combination the above frequency bands anti-stealth and anti-stealth networking, and through its working characteristics, enhancement of the anti-stealth effect.

First, the system can work at multiple frequency bands at the same time, and only at certain frequencies can it effectively reduce the target stealth materials of the RCS, and can take advantage of the VHF/UHF and HF bands. In fact, in the spectrum of radio and television signals across the electromagnetic environment, the network against radar reconnaissance system can take advantage of these signals, in a non-cooperative source working methodology, to detect the stealth target. In the absence of radiation sources to use, it can transmit a signal source in the spectrum to illuminate the stealth target. Through multiple station data fusion, the system can solve the problem which has measuring accuracy and resolution that are too low for meter wave radar, and the problem of low positioning precision of the over-the-horizon radar.

Second, in respect of anti-stealth capability, the system is better than modern netted radar technology. The anti-stealth ability of netted radar is joined with anti-stealth radar, such as the meter wave radar mentioned earlier, over-the-horizon radar, millimeter wave radar, and impulse radar. But the detection ability of netted radar is subject to single radar detection ability, and it is difficult to greatly improve on the position precision. Such as for long-range stealth target detection, the detection accuracy basically depends on the over-the-horizon radar positioning accuracy of the network. At close range, the use of meter wave radar and the need for the combined action of multiple kinds of the same radar can be improved through the data fusion technology The cost is much more expensive than the network against radar reconnaissance system, so for the construction of an air defense system, network radar countermeasure system deployment is more effective and practical.

6.2.3 Sending and Receiving Allocation and Data Sharing Improve Anti-stealth Capability

Because stealth technology, especially plastic technology, cannot reduce a stealth aircraft's RCS in all directions, therefore, the radar needs to avoid the RCS stealth aircraft reduced obviously direction, from another direction to illuminate the stealth aircraft. It is possible to keep the original range of the stealth target detection ability, and it is the base of the network radar countermeasure reconnaissance system that makes use of sending and receiving characteristics to resist shapes designed with stealth technology. Due to the transceiver separation system, the network radar countermeasure reconnaissance system can make full use of the lateral area of the radar's target scattering in dealing with stealth targets, as well as forward scattering and target Doppler beat frequency detection.

1. The use of stealth target lateral scattering energy

   The starting point of stealth aircraft design is to use plastic technology, impedance load to lead partial, cheat, or increase the tracking error to the mono-static radar. It can make the mono-static radar difficult to conduct continuous observation and measurement, and can only obtain some irregular discontinuity plots. Make the radar wave scattering deflection in the other direction, and make sure it has a tiny RCS within the aircraft nose cone in the $\pm 30°$ direction. Only some key parts are difficult to change shape, through the absorption wave structure used on the parts of material and absorbing coatings. So the ability of a stealth target's lateral scattering does not reduce and, in some directions, actually increases. A mono-static radar cannot receive these lateral scatterings, but the system receiver can. For stealth targets, the strong scattering direction of the beam is narrow. Because the system has multiple receiver and transmitter stations, we can use the information of these multiple receives and transmitters for target detection, making full use of the lateral scattering signal of the stealth target. In addition, the lateral scattering of stealth targets is less than for normal targets, but they still have considerable RCS. Therefore, for the system make a reasonable disposition, reasonable selection of the distance between stations, forming a right angle, through data sharing and data fusion technology, can improve the effectiveness of the stealth target detection and tracking performance.

2. The use of stealth target forward scattering energy

As is widely known, whether it is metal or coated with absorbing material, the target of forward scattering is:

$$\sigma_F = 4\pi A^2 / \lambda^2 \tag{6.19}$$

In the formula, σ_F is the forward scattering RCS of the target; A is the projection area of the target, and λ is radar working wavelength.

For example, for $D = 1.5$ m diameter ball, when $\lambda = 0.1$ m, the forward scattering RCS is 4000 m^2 and the sphere's backward scattering RCS is only 1.77 m^2. But the target forward scattering of the main lobe is also very narrow, so to make a reasonable use of the forward scattering, many factors also need to be considered in system deployment and in battle, such as tactics and the rational allocation of sites, which can be more comprehensive uses and advantages of data sharing, to receive the forward scattering energy of the stealth target, so as to avoid the angle blind area, range blind area, and Doppler blind area of the stealth target, improving the system's detection of stealth targets and tracking performance.

3. The use of the Doppler beat frequency

On the baseline of transmitting and receiving stations, the target has very strong scattering. When detecting and tracking the stealth target, measures should be taken to make full use of the strong scattering energy. During the initial period of bi-static radar development, people found aircrafts and ships change according to the target through the baseline creative of the Doppler beat frequency. Regardless from which direction the target crosses the bi-static radar baseline, the Doppler beat frequency changes to zero when it is across the baseline. The beat frequency receiver can detect this zero beat frequency, note down the moment, and then, through multi-point prediction, track the target. The network against radar reconnaissance system with multiple extra work characteristics, with simple 3XT 5XR system as an example, the system can form 15 baselines. When the target crosses the system's coverage, then crosses between more densely packed baselines, through timely records of the zero beat frequency time, under the condition of data sharing in the whole system and through calculation, the target track measurement can be realized.

In conclusion, the network against radar reconnaissance system, in respect to anti-stealth, has all the advantages of the bi-static radar, but at the same time overcoming the bi-static radar's main problems in target detection cannot, such as blind area, and improve the detection performance of radar. Compared with the mono-static netted radar, the netted radar makes use of anti-stealth bi-static (multi-) radar base to solve the problem of stealth target detection. In fact, to increase bi-static (multi-) radar in a radar network to implement the radar stealth target detection, the performance is limited by the nature of bi-static (multi-) base radar detection performance. Therefore, the network against radar reconnaissance system, in respect of improvement of anti-stealth capacity, was better than the mono-static radar network.

6.3   Anti-radiation Attack Performance Analysis of a Network Radar Countermeasure System

Destroying anti-radiation is mainly used for anti-radiation missiles (ARM) and anti-radiation UAVs (ARD), used as the representatives of anti-radiation weapons. For radar radiation direct firepower, it is hard to kill any means of electronic warfare. They are using passive radar seeker guided warheads to attack the radiation source, which has become an important weapon in the modern war to gain mastery of the air, and has become indispensable.

In the Vietnam War, the U.S. Air Force used the AGM-45 first generation of anti-radiation missiles, which were used to destroy the anti-aircraft artillery of anti-aircraft guns and anti-aircraft missile guidance radar, and the effect was amazing. In the Middle East War, Israel used the AGM-45 anti-radiation missile to destroy the Soviet system SA-2, SA-3 ground-to-air system with a high hit rate. The U.S. Air Force stocked against Libya the use of the AGM-88A high-speed anti-radiation missiles, which destroyed the Libyan air defense radar, and Libya's air defense radar suffered heavy losses. In the Gulf War, the U.S. launched a variety of anti-radiation missiles, destroying 60 % of Iraq's air defense radar, causing a 95 % loss of battle effectiveness, thus resulting in Iraq's air defense system being paralyzed and causing significant loss and defeat. Many examples illustrate a serious threat of anti-radiation missile radar and its major influence on modern warfare.

6.3.1   The Advantage of the Network Radar against Anti-surveillance System and Anti-radiation Weapons

The modern anti-radiation weapon has the characteristics of high sensitivity, high speed, and great intelligence. Indeed, it is difficult to deal with the existing single base radar threat. But the anti-radiation weapon has its limitations, and according to these limitations, ways can be found to deal with them. The limitations of anti-radiation weapons are mainly reflected as follows. (1) The dependence on the radiation source. Anti-radiation weapons use passive seeker guidance, and can only attack the radiation source, so they can only attack the transmitter of the single base radar and other transmitters of radar systems, and for a receiver, it is powerless. (2) Seeker sensitivity is limited. Because anti-radiation weapons have the ability to attack the RF agile radar, its seeker instantaneous bandwidth is large (about 500 MHz). So, in defining the sensitivity of the seeker, its role is limited distance, usually $20 \sim 50$ km, and it is not yet over 100 km. (3) Signal recognition ability is limited. Because of the limitation of volume, anti-radiation weapons can only load a limited number of priority radar parameters into the database, and it is impossible to include all radars, especially radars which are concealed in peacetime. Once the anti-radiation weapon launch radar suddenly reaches a state of no parameters in the database, the weapon will lose track. (4) Anti-radiation weapons cannot work on low frequency. Because of the volume restrictions, anti-radiation weapons cannot accommodate low-frequency antenna, so there is still no anti-radiation weapon operating in the UHF or VHF bands.

The working characteristics of the network against radar reconnaissance system itself determine its outstanding ability when dealing with anti-radiation weapons. First of all, the concealment of the passive mode system destroyed the anti-radiation weapons dependence on the radiation source, without providing radiation targets for the anti-radiation weapons. Second, the separation system of sending and receiving for radar anti-radiation weapons provides further guarantees. In the tactical configuration, it is possible to configure the transmitting station at the rear, which is away from the forefront of battle, at a distance of about 100 km. This is beyond the sensitivity range of the modern anti-radiation weapon seeker, and anti-radiation weapons cannot attack the transmitter. The single base radar network and its system performance are dependent on the single radars in play, and for a single base radar, detection range is certain. In order to guarantee the network coverage of the air defense radar system, the radar cannot be placed too far back, so it may not be beyond the scope of an anti-radiation weapons attack.

In addition, anti-radiation weapons use only a system of the transmitting station, and the system is generally equipped with multiple transmitting and receiving stations. Even if the system is under attack and the transmitting station is destroyed, then the receiving station may continue to operate, and the system will not cause fatal injuries. For a single radar, when subjected to an anti-radiation weapons attack, the entire radar is destroyed and will fail to work. When the single radar networking is being attacked, the entire radar can be destroyed. When a single base radar networking is being attacked, it can continue to operate until all radars are destroyed.

6.3.2   Anti-destroying Capability Analysis

Considering the anti-destroying capacity of the network radar countermeasure system, studies have shown that the probability of a single radar's ability to resist being destroyed is expressed as:

$$P_{k0} = 1 - P_m P_a P_b \qquad (6.20)$$

P_m is the probability of a single radar discovering an anti-radiation weapon; P_a is anti-radiation weapons found after the radar, i.e., the probability of attacks being implemented; and P_b is the probability of anti-radiation weapons hitting the radar.

In the bi-static double/multiple radar in pairs/base radar transceiver division as well as radar reconnaissance system, its ability to resist to destroy is equivalent to the transmitting station's ability to resist destruction. For the bi-static radar, the station platform, usually at the rear or in the air, it is found that for anti-radiation weapons, the attack probability and the hitting probability are lower than those of single radar base. This resulted in the probability of anti-radiation weapons destroying the bi-static radar being better than that of the single base radar. Typical data show that: for single base radar, the anti-destroy ability is:

$$P_{k0} = 1 - 0.8 \times 0.85 \times 0.8 = 0.456 \qquad (6.21)$$

For the bi-static radar, the anti-destroy ability is:

$$P_{k1} = 1 - 0.4 \times 0.8 \times 0.6 = 0.808 \qquad (6.22)$$

The network radar countermeasure system problem is much more complex. First, set a more optimized network reconnaissance radar countermeasure system configuration, in order to reduce the risk of a transmitting station being attacked, it will launch after the station in order to improve the detection distance, and the need have the station at the front. Figure 6.2 shows the configuration diagram. The figure only shows the location of the receiving station and the transmitting station configuration; there is no detail on the hubs and network structures. According to the working principles of the system, as long as the transmitting station has not been completely destroyed, with a few of them even missing a transmission job, the system still has some ability to work. So, in the analysis of the anti-destroy ability, the probability analysis shows the transmitting station was destroyed.

In the calculation, we need to consider two factors: the power of the anti-radiation weapon radius and the distance between the transmitter. First, consider a system equipped with two stations. The basic configuration model is shown in Fig. 6.3.

In the figure, (a) is the diagram of the destroyed anti-radiation weapons, which shows two transmitting stations in the case of independent events and (b) represent the case of the destruction of the two transmission stations for non-independent

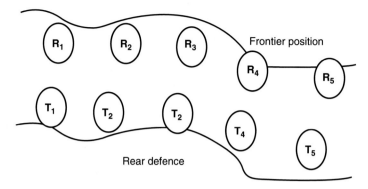

Fig. 6.2 Reconnaissance 去掉

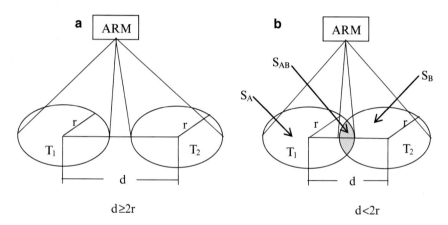

Fig. 6.3 Configuration showing the relationship between two transmitting stations

events. Actually, (a) is a special case diagram of (b). For (b), the figure shows the model where transmitting station T1 was destroyed with probability A, and the probability of transmitting station T2 being destroyed is B, then the anti-destroy ability of the whole system is:

$$P_{k2} = 1 - P(AB) - 2P(A \cap \overline{B})P(B) \tag{6.23}$$

wherein $P(AB)$ is the probability that the two transmission stations were destroyed at the same time and $2P(A \cap \overline{B})P(B)$ is the probability that the two transmitting stations are destroyed. In formula (6.22), on the basis of the bi-static radar anti-destroy probability, and assuming that the placement of missiles in the region is evenly distributed, we obtain:

$$P(A) = P(B) = (1/2)(1 - P_{k1}) = (1/2)P \tag{6.24}$$

In the formula, P is the probability that the bi-static radar was destroyed. According to formula (6.22) $P = 0.192$. From combining formulas (6.23) and (6.24), we have:

$$P_{k2} = 1 - \frac{S_{AB}}{2S_A}P - \frac{1}{2}\left[1 - \frac{S_{AB}}{2S_A}\right]P^2 \tag{6.25}$$

In the formula, S_A, S_B, and S_{AB} are the illustrated areas. Taking $d = 300$ m, $r = 200$ m, and $P = 0.192$, based on the type of anti-destroy, the coefficient obtained in the case of two transmitting stations is 0.9704.

For three transmitting stations, there are many possibly ways to deploy them. For example, it can have a linear distribution station, or a triangle station can be divided into overlapping and non-overlapping various cloth station circumstances, such as that shown in Fig. 6.4. For more transmitting stations, its distribution station mode is even more complex, in accordance with the operational needs determined by specific programs. In a variety of fabrics that stand against the way of destruction, the system capacity will be different. For simplicity, we only discussed the distribution of a linear interval of three stations. Figure 6.5 shows the parameters in this case. By the use of the previous method, the anti-destroy coefficient of the system comprising three transmitting stations as follows similar to the coefficient can be obtained as follows:

$$P_{k3} = 1 - \frac{2}{9}\frac{S_{AB}}{S_A}\left[2 - \frac{S_{AB}}{S_A}\right]P^2 - \frac{4}{27}\left[\left[1 - \frac{S_{AB}}{S_A}\right]\left[2 - \frac{S_{AB}}{S_A}\right] - \frac{1}{2}\right]P^3 \tag{6.26}$$

The meaning of the formula parameters and formula (6.25) shows that the same formula will promote more transmitting stations that can be obtained under the network radar countermeasure system against the anti-destroying factor.

In order to compare the network of radar surveillance system and a single base against networking destroy anti-radar capabilities, previous conclusions can be used to calculate the corresponding anti-destroying factor with the parameters set to: $d = 300$ m, $r = 200$ m, and $P = 0.192$. For a single radar network in accordance with the same configuration parameters, the results are shown in Table 6.1, where n is the number of transmitting stations. As can be seen from the data in the table, the anti-destruction capacity of the network of radar reconnaissance system is superior to that of a single radar network.

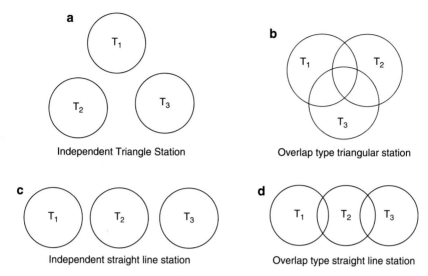

Fig. 6.4 Three launching stations arrangement diagram

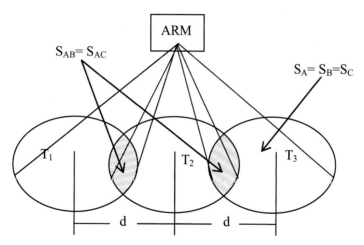

Fig. 6.5 Schematic showing the destruction probability of three transmitting stations under anti-radiation weapons attack

Table 6.1 Comparison the anti-radiation attack capability of the network radar countermeasure system and a single base radar network

n	1	2	3	4
Network radar countermeasure system	0.8080	0.9816	0.9984	0.9999
Single base radar network	0.4536	0.8520	0.9642	0.9918

6.4   Network Radar Countermeasure System Against Anti-Low-Altitude Penetration Performance Analysis

Low-altitude penetration is the use of the Earth's curvature and altitude changes of the terrain to block and surface-to-air screen for the blind area of air defense facilities, and the use of anti-aircraft weapons requires scheduled time and other favorable conditions. The aircraft, missiles, and others and carry out fast covert surprise attacks in hostile zones and complete the task before the counterattacks and withdrawal of the enemy, before air defense weapons systems are able to respond. Low-altitude penetration radar blinds greatly reduced the probability of aircraft being detected and, therefore, the probability of being destroyed, thereby increasing the viability and effectiveness of their mission.

In order to implement low-altitude penetration effectively, most armies have modern weapons flying at low altitude and with high maneuvering performance, as well as high-precision navigation performance, automatic, real-time, realistic terrain display capabilities, terrain following, terrain avoidance, terrain collision, threat avoidance, the ability to work day and night and all-weather capability, good anti-jamming, and hidden properties.

Development of low-flying weapons and air defense systems pose a serious threat. The problems of modern radar in detecting low-altitude targets encountered include: (1) reflection from the ground or sea clutter jamming is extremely strong, and the useful echo is completely submerged; (2) limited curvature of the Earth; (3) hidden topography; (4) jamming ground (sea) reflected waves caused by low elevation blind.

Flexibility and networking of the network reconnaissance radar countermeasure system improves the defense capabilities of the low-altitude reconnaissance system to provide a guarantee. Its form is not restricted according to the operational requirements for the deployment of network elements, and, so, in the case of the low-altitude defense mission being clear, can give priority to effectiveness at a low-altitude station arrangement. In addition, the timely response capability of the system network into the system provides a reliable guarantee. Below is a performance analysis of the network against radar reconnaissance system when dealing with the low-altitude penetration goals to consider problems and responses.

6.4.1   Bi-static Radar Increasing the Detection Range

Due to the limitations of linear propagation of electromagnetic waves, the Earth can only be found in a certain target line of sight. That is to say, flying higher than the target "radar horizon" means that flight will not be discovered. The lower the target to be observed, the closer to the horizon. Therefore, increasing the height of the radar antenna extends the radar range. We set the radar line of sight R (km) and radar antenna height h (m), as shown in Fig. 6.6.

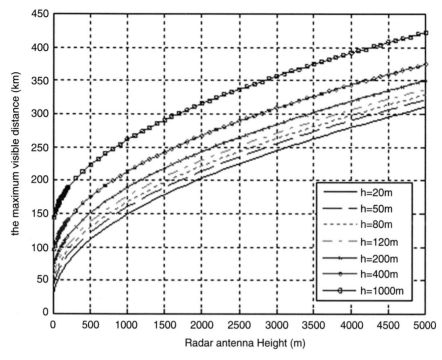

Fig. 6.6 Radar antenna height and target height relations in different situations and visible distances

For the target flying height H (m), then the relationship is given by the expression:

$$R = 4.12\left(\sqrt{h} + \sqrt{H}\right)(km) \tag{6.27}$$

The first term is a coefficient to consider in normal refraction. According to the formula, it is possible to draw a relationship between the radar range radar antenna height and the target's flying height, as shown in Fig. 6.6. In the figure, the curve gives the maximum range of the radar antenna and the time between targets detection at different flight altitudes. For a single radar, without considering the signal power and receiver sensitivity, the maximum effective distance and visible distance are equivalent. For a single radar, dual diffraction exists in the electromagnetic wave propagation conditions, and attenuation and diffraction loss are large, so its effect is limited by distance; For the bi-static radar, since the transmission and reception separator can be a receiving station before the match, it is possible to reduce the electromagnetic wave diffraction from the target, in terms of the transmitting station, by the diffraction of the target irradiated by direct vision wave station the received echo signal, only one way of diffraction attenuation, to a certain extent, which improves the distance of radar detection of low-altitude targets.

For low-altitude target detection, we need to configure the receiving station of the network of radar reconnaissance system at the front, both in line with the network of radar reconnaissance system against attacks against ARM needs, but also to improve the low-altitude target detection range of the system. Consider a radar network with three transmitting stations and receiving stations consisting of five combat reconnaissance systems. The minimum systems to calculate its base with a single network under the same conditions for the radar detection range of low-altitude targets are compared to the results against the network reconnaissance radar system advantages in terms of low-altitude defense. In order to facilitate comparison, consider the range of 360° in the biggest anti-aircraft range. On the network performance of the radar to a three identical single-blind radar base complement formation, the best way is a cloth station equilateral triangle station. Now order now a false radar detection range of 50 km. In order to make the maximum detection range, make it stop when the cloth distance between stations is 100 km, then the detection range of the network after is shown by the dashed line in Fig. 6.7, and the visible maximum detection distance in all directions is probably about 110 km. The networking radar station on the same mode of distribution, for which the netted radar transmitting station is configured to position the station by a peripheral distribution station, the solid line in Fig. 6.7 shows the front with six stations with a way to form a network radar countermeasure system fight, this time against the network of radar surveillance system to detect the orientation in a full scope significantly increased the maximum detection range of up to 140 km. Here, calculation does not consider the use of the case with the case under the former stations unidirectional wave propagation diffraction. In consideration of the diffraction detection range, the detection range of the network against the reconnaissance radar system will be further expanded. Therefore, it is certain that for the network against the reconnaissance radar, the networking capability to detect low-altitude targets have significantly improved compared to a single radar system under the same conditions. Of course, in actual use, we can focus on specific targets in the direction of the defense, for the proper deployment location of the transceiver station, so that the resources of the system are utilized to the fullest.

6.4.2   Passive Detection Improves the Low-Altitude Target Detection Capability

Carrying out low-flying low-altitude targets usually requires a radar altimeter for a topographic match, such as for cruise missiles. So, it can make full use of the passive mode of the network radar countermeasure system to complete the

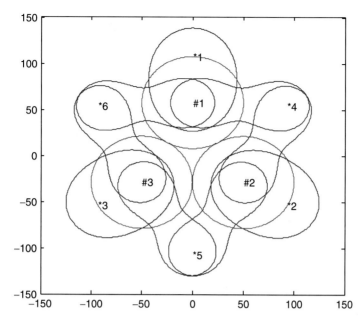

Fig. 6.7 The schematic diagram of Networking Radar Systems and Networks Radar Countermeasure System against low-altitude radar detection range

detection of low-altitude targets. In this case, the detection performance depends on the location of the radar receiver and target height relationship without considering the case of electromagnetic diffraction, and the detection range can be calculated according to (6.27). When smooth ground or sea waves have a smaller amplitude, it will produce secular reflection, electromagnetic wave propagation mode, and radar target geometry as shown in Fig. 6.8. The figure shows that a radar signal can take two routes to the radar receiver: a radar signal directly into the radar receiver or be reflected by the ground (or sea surface) into the radar receiver, thus forming multi-path signals in the receiver.

As for the question of reducing the impact of multiple paths on the radar to detect and track low-angle performance, currently, there are single radar narrow beam antenna designs, double zero designs, off-axis tracking technology, frequency diversity, spatial filtering, asymmetrical antenna pattern and so on. In the software design considerations, enhanced signal processing and data processing capabilities, according to different sea models, multi-objective valuation method, and different data processing methods can be used. The network reconnaissance radar countermeasure system can also consider using a narrow beam antenna design, reducing the probability of fission of the main lobe.

But, more importantly, the biggest advantage of the network of radar reconnaissance system is the distribution of multiple stations and multi-station data, for information fusion. When a task requires a regional focus to detect low-altitude

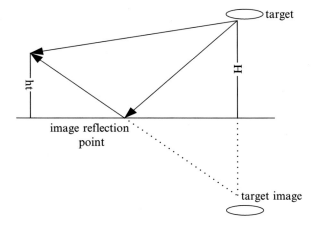

Fig. 6.8 Secular reflection geometric relation under the passive mode

targets, in addition to taking the reconnaissance antenna elevation, we can also take advantage of the network for data sharing of the radar reconnaissance system and its powerful data fusion, data for a plurality of receiving stations in time for the integration process, to achieve effective low-altitude target detection and location.

Bibliography

1. 姜秋喜, 王希勤, 丁锋, 李明亮. 网络雷达及其定位精度分析 [C]. 电子对抗系统专业委员会与雷达系统专业委员会联合学术年会, 昆明, 2006.10.
2. 姜秋喜, 李明亮, 丁锋. 网络雷达及其MIMO状态分析 [C]. 中国电子学会电子对抗分会第十五届学术年会论文集, 扬州, 2007.10: 377–379.
3. 杨振起, 张永顺, 骆永军. 双(多)基地雷达系统 [M]. 国防工业出版社, 1998.
4. 谢卓译,张直中等校, M. I. Skolnik. 雷达手册(第二版) [M]. 国防工业/电子工业出版社, 1978/2003.7.
5. 保铮, 张庆文.一种新型的米波雷达-综合脉冲与孔径雷达 [J],现代雷达, Vol. 17, No. 1, 1995:1–13.
6. 陈永光, 李修和, 沈阳.组网雷达作战能力分析与评估 [M]. 国防工业出版社, 2006.
7. 朱旭东. 一种有源和无源综合探测系统的研究 [J]. 现代雷达, Vol. 25, No. 6, 2003: 8–11.
8. 张格学. 雷达对抗战术 [M]. 解放军电子工程学院, 2002.7.
9. E. Fishler, A. Haimovich, R. Blum, D. Chizhik, L. Cimini, R. Valenzuela. MIMO radar: an idea whose time has come [C]. Proceedings of the 2004 IEEE Radar Conference, Philadelphia, PA, USA, 2004: 71–78.
10. E. Fishler, A. Haimovich, R. Blum, L. Cimini, D. Chizhik, R. Valenzuela. Performance of MIMO radar systems: Advantages of angular diversity [C]. Conference Record of the Thirty-Eighth Asilomar Conference on Signals, Systems & Computers, Vol. 1, 2004:305–309.
11. E. Fishler, A. Haimovich, R. Blum, L. Cimini, D. Chizhik, R. Valenzuela. Spatial diversity in radars-models and detection performance [J]. IEEE Transactions on Signal Processing, vol. 54, No. 3, 2006:823–838.
12. 杨小牛, 陆安南, 金飚. 宽带数字接收机 [M]. 电子工业出版社, 2002.10.
13. [美]Richard G. Wiley著,吕跃广等译. 电子情报 (ELINT)——雷达信号截获与分析 [M] 千京, 电子工业出版社, 2008.3.
14. 沈凤麟, 叶中付, 钱玉美. 信号统计分析与处理 [M]. 中国科学技术大学出版社, 2003.
15. 丁鹭飞, 耿富录. 雷达原理(第三版) [M]. 西安电子科技大学出版社, 2002.
16. P.van Genderen and W.J.H. Meijer. Non Coherent Integration in a Medium PRF Radar [C]. IEEE International radar conference, 1995: 91–93.
17. Y. Teng, H.D. Griffiths, C.J. Baker. Netted radar sensitivity and ambiguity [J]. IET Radar Sonar Navig, 1,(6), 2007:479–486.
18. T. Tsao, M. Slamani, P. Varshney, D. Weiner, H. Schwarzlander. Ambiguity function for a Bistatic Radar [J]. IEEE Trans. Aerospace and Electronic Systems, Vol. 33, No. 3, 1997:1041–1051
19. 汪学刚, 张明友. 现代信号理论 [M]. 千京, 电子工业出版社, 2005: 204–210.
20. G. San Antonio, D.R. Fuhrmann, F.C. Robey. MIMO Radar ambiguity functions [J]. IEEE Journal of Selected Topics in Signal Processing, vol. 1, No. 1, 2007:167–177.
21. Qu JinYou, Zhang JianYun, Liu Chun Quan. The Ambiguity Function of MIMO Radar [C]. IEEE 2007 International Symposium on Microwave, Antenna, Propagation, and EMC Technologies For Wireless Communications, 2007:265–268.

© National Defense Industry Press, Beijing and Springer-Verlag Berlin Heidelberg 2016   299
Q. Jiang, *Network Radar Countermeasure Systems*,
DOI 10.1007/978-3-662-48471-5

22. 刘同明, 夏祖勋, 解洪成. 数据融合技术及其应用 [M]. 国防工业出版社, 1998.

23. 赵宗贵, 耿立贤, 周中元等编译. 多传感器数据融合 [M]. 电子二十八所, 1993.

24. Lawrence A, Klein. Sensor and data fusion concepts and applications [J]. SPIE Optical Engineering Press, Washington, 1999.

25. Dall. L. Mathematical techniques in multisensor data fusion [M]. Artech House, 1992.

26. Carl B Frankel, Mark D Bedworth. Control, estimation and abstraction in fusion architectures: Lessons from human information processing [C]. Proceedings of 2000 International Conference on Information Fusion. Paris, 2000, 130–137.

27. Kokar M, Kim K. Review of Multisensor Data Fusion Architectures and Techniques [C]. Proceedings of the International Symposium on Intelligence Control, Chicago, August, 1993, 261–266.

28. Vessels P A. Service-based architecture for the network-centric battlefield [C]. Proceedings of SPIE, 2002.

29. 孙尧, 朱林, 徐兴杰, 张晓因. 基于数据融合树的C3I信息融合系统体系结构设计 [J]. 航空学报, Vo 27, No. 2, 2006: 305–309.

30. Alan N. Steinberg, Christopher L. Bowman, E. white. Revisions to the JDL Data Fusion Model [C]. Proceedings of SPIE Volume 3719, Sensor Fusion: Architectures, Algorithms and Applications III, 1999 Bellingham USA. 430–441.

31. Christopher L. Bowman. The data fusion tree paradigm and its dual [C]. Proc. 7th National Symposium on Sensor Fusion. 1994.

32. Christopher L. Bowman. Data integration (fusion) tree paradigm [C]. Proc. SPIE Vol. 1698 Signal and Data Processing of Small Targets, 1992:372–381.

33. 丁锋, 姜秋喜, 李明亮. 网络雷达及其数据融合问题研究 [J]. 中国电子科学研究院学报, 第3期, 2008.6:285–292.

34. 孙仲康, 周一宇, 何黎星. 单多基地有源无源定位技术 [M]. 千京:国防工业出版社, 1996

35. 张兴如. 无源定位及其性能分析 [D]. 成都电子科技大学硕士毕业论文, 2001.

36. 王成,李少洪,黄槐. 测时差定位系统定位精度分析与最优布站 [J]. 火控雷达技术, 2003, 32(1):1–6.

37. Singer R A. Estimating optimal tracking filter performance for manned maneuvering targets [J]. IEEE Transactions on Automatic Control, 1970, 20(6):473–383.

38. X. Rong Li, Survey of Maneuvering Target Tracking [J]. Part I: Dynamic Models, IEEE Transactions on Aerospace and Electronic Systems, 2003, 39(4):1337–1364.

39. 韩崇昭, 朱洪艳, 段战胜等. 多源信息融合 [M], 清华大学出版社, 2006.4.

40. E. mazor, J. Dayan, Y. Bar-Shalom. Interacting Multiple Model in Target Tracking [J]. IEEE Transactions on Aerospace and Electronics, 1998:103–124.

41. D. Wllner, C.B. Chang, and K.P. Dunn. Kalman Filter Configurations for multiple Radar Systems, M.I.T. Lincoln Laboratory Report No. TN-1976-21(14 April, 1976).

42. Gan Qiang, Harris C J. Comparison of Two Measurement Fusion Methods for Kalman-filter-based Multisensor Data Fusion [J]. IEEE Transactions on Aerospace and Electronic Systems, 2001, 37(1):273–280.

43. Krieg M L., Gray D A. Radar and Optical Track Fusion Using Real Data [C]. Proc. 1st Australian Data Fusion Symposium(ADFS-96), Adelaide, Australia:25–30.

44. 刘同明等. 数据融合技术及其应用 [M]. 千京:国防工业出版社, 1998.

45. 刘刚. 多目标跟踪算法及实现研究 [D], 博士论文, 西千工业大学, 2003, 7.

46. R.J. Fitzgerald. Development of Practical PDA Logic for Multitarget Tracking by Microprocessors, In Multitarget-Multisensor Tracking: Advanced Applications, (ed. B.-S.Y.) [M], Norwood, MA, Artech House, 1990.

47. J.A. Roecker, G.L. Phillis. Suboptimal joint probabilistic data association [J], IEEE Trans on AES, 29(2), 1993: 510–517.

48. L.Y. Pao, C.V. Frei. A comparison of Parallel and Sequential Implementation of a Multisensors Multitarget Tracking Algorithm. Proc. 1995 American Control Conf. Seattle Washington, June 1995:1683–1687.

49. L.Y.Pao. Centralized Multisensor Fusion Algorithm for Tracking Applications. Control Engineering Practice, 2(5), Oct. 1994:875–887.

50. 周宏仁, 敬忠良, 王培德. 机动目标跟踪 [M], 千京, 国防工业出版社, 1991.

51. Y. Bar-Shalom, E. Tse. Tracking in a cluttered environment with probabilistic data association [J], Automatica, 11(9), 1975: 451–460.

52. Y. Bar-Shalom, T.E. Fortmann, M. Scheffe. Joint probabilistic data association for multiple targets in clutter [C], Proc. 1980 Conf. Information Sciences and Systems, Princeton Univ., 1980.

53. 王杰贵, 罗景青. 基于多目标多特征信息融合数据关联的无源跟踪方法 [J]. 电子学报, 2004 (6): 1013–1016.

54. 付耀文. 雷达目标融合识别研究 [D]. 国防科技大学工学博士论文, 2003.

55. 黎湘, 付耀文, 庄钊文等. 目标识别决策层融合神经网络算法研究 [J]. 电子科学学刊, 2000(22): 692–696.

56. Lorenz F P, Biermann J. knowledge-based fusion of formats: discussion of an example [C]. Proc. of the Fifth International Conference on Information Fusion, 2002(1):374–379.

57. Smith J F. Fuzzy logic resource manager: multi-agent fuzzy rules, self-organization and validation [C]. Proc. of the Fifth International Conference on Information Fusion, 2002(1): 199–206.

58. 姜秋喜, 董晖. 基于单个脉冲的雷达个体识别技术研究 [J]. 舰船电子对抗, Vol. 29, No. 4, 2006: 60–67.

59. 罗景青等.雷达对抗原理 [M]. 解放军出版社, 2003.

60. 李洪兴. 工程模糊数学方法及应用 [M]. 天津科学技术出版社, 1991.

61. 王杰贵, 靳学明, 罗景青. 基于ESM与ELINT信息融合的机载辐射源识别 [J]. 电子学报, Vol. 34, No. 3, 2006:424–428.

62. 魏巍贤,冯佳.多目标权系数的组合赋值方法研究 [J]. 系统工程与电子技术, 1998年第2期, 14-16

63. 罗志增,叶明.用证据理论实现相关信息的融合 [J]. 电子与信息学报, Vol. 23, No. 10, 970-974, 2001

64. Hongwei Zhu, Otman, Basit. A Scheme for Constructing Evidence Structure in Dempster-Shafer Evidence Theory for Data Fusion [C]. Proceedings 2003 IEEE International Symposium on Computational Intelligence in Robotics and Automation, Japan, 2003, 16–20.

65. 潘继飞, 姜秋喜, 毕大平. 雷达"指纹"参数选取 [J]. 现代防御技术, Vol. 35, No. 1, 2007:75–79.

66. 赵国庆.雷达对抗原理 [M]. 西安电子科技大学出版社, 1999.

67. D.W. Allan. Statistics of Atomic Frequency Standards, Proc. IEEE 54, 221–230, Feb, 1996.

68. 刘海燕, 赵宗贵, 巴宏欣. 一种基于加权证据合成多传感器目标识别方法 [J]. 解放军理工大学学报, Vol. l6, No. 6, 2005:521–524.

69. Capelle A S, Maloigne C F, Colot O. Introduction of spatial information within the context of evidence theory [C]. IEEE International Conference on Acoustics, Speech, and Signal Processing. 2003(2): 785–788.

70. 戴亚平等.多传感器数据融合理论及其应用 [M]. 千京理工大学出版社, 2004.

71. 何友,陆大瑈,彭应宁.多目标多传感器模糊双门限航迹相关算法 [J]. 电子学报, 1998年第3期, 15-19.

72. 何友,彭应宁,王国宏.基于模糊综合函数的航迹关联算法 [J]. 电子科学学刊, 1999年第1期, 91-96.

73. 何友,黄晓东.基于模糊综合决策的航迹相关算法 [J]. 海军工程大学学报, 1999年第4期, 1-11.

74. 何友,黄晓东.分布式多因素模糊综合评判航迹相关算法 [C]. 第七届全国雷达年会论文, 417–420, 1999.

75. MEI Wei, Shan Gan-lin, Zhou Yun-feng. Theoretical performance of a multiscan track-to-track association algorithm [J]. Signal Processing. 2005, 85 (1):15–22

76. Hong L. Cui N Z. An Interacting Multipattern Probabilistic Data Association (IMP-PDA) Algorithm for Target Tracking [J]. IEEE Trans. On Automatic Control (S0018-9286), 2001, 46(8): 1223–1236

77. 张池平,王晓明,崔祜涛.D-S证据理论在航迹关联问题中的应用　[J].　传感器技术Vol. 24, No. 8, 71-74, 2005

78. 张锡熊. 21世纪雷达的"四抗" [J]. 雷达科学与技术, Vol. 1(1), 2003.06:1–6.

79. 熊少华, 钟持瑞. 强噪声干扰环境中的雷达无源定位研究 [J]. 现代雷达, Vol. 22(3), 2000.02: 17–19.

80. 李可达. 现代雷达基本抗干扰技术 [J]. 航天电子对抗, 2004(2): 15–19.

81. 郭克成, 陆静. 对双基地雷达的抗干扰能力及有效干扰区分析 [J]. 现代雷达, Vol. 26(9), 2004.09: 20–23.

82. 朱华邦, 杜娟. 雷达抗干扰技术的新特点及发展方向 [J]. 飞航导弹, 2004(5): 52–54.

83. 羊爱国, 杜伟庆, 赵纯锋等. 单、双基地雷达四抗能力比较 [J]. 舰船电子对抗, Vol. 26(5), 2003.05: 29–31.

84. 吴明忠,赵振声,何华辉. 隐身与反隐身技术的现状与发展 [J]. 上海航天, 1996(3):36–42.

85. 蒋庆全. 反隐身雷达技术发展探析 [J]. 火控雷达技术, Vol. 33(6), 2004.06: 28–32.

86. 薛晓春,王学华.　隐身与反隐身技术的发展研究　[J].　现代防御技术, 　Vol.　32(2), 2004:60–65.

87. 焦国娥. 反辐射武器对雷达的威胁及其对策 [J]. 舰船电子对抗, 1998(2): 26–29.

88. 程柏林,姜永金,刘捷等. 多基地雷达抗摧毁效能分析 [J]. 航天电子对抗, 2001(3):10–12.

89. 蒋谱成, 刘英芝. 雷达抗反辐射导弹的方法概述 [J]. 舰船电子对抗, 2001(2):6–9.

90. 雷旺敏, 张飚. 岸基雷达的抗海杂波措施 [J]. 现代雷达, Vol. 28(5), 2006.05:5–8.

91. 陈振邦. 低空防御与低空补盲雷达的发展 [J]. 现代防御技术, 1997(3):16–23.

92. 王英钧. 先进的低空突防技术 [J]. 航空电子技术, 1995(1):35–41.

93. 周京杭, 任渊. 海面异常杂波的特征及抑制方法探讨 [J]. 雷达与对抗, 2005(4):14–17.

94. 赵巨波, 符燕, 耿文东. 海杂波统计特性分析 [J]. 现代雷达, Vol. 27(11), 2005:4–6.

95. TennenhOuse D L, Snllth J M, Sincoske W D, et al. A survey of active network research [J]. IEEE Common. Mag., 1997, 35(1):80–85.

96. Nonnenmacher J, Biersack E W, Towsley D. Parity- based loss recovery for reliable multicast transmission [J]. IEEE ACM Trans. on Networking, 1998, 6(4):349–361.

97. Tanenbaum, Andrew S. Computer networks (Fourth edition) [M]. Prentice Hall, 2002.

98. Comer, D., Computer Networks and Internets [M]. Prentice Hall, 2001.

99. 李冬霞,苏广川. 组网雷达系统的动态管理技术研究 [J]. 系统工程与电子技术, 2001, 23 (4):58–60.

100. 朱丽莉, 王朝墀. 制导雷达组网技术研究 [J]. 现代雷达, 2003, 25(7):1–3.

101. 马晓岩,谢跃权,王立振. 分布式遥控雷达中基于宽带网的信息传输技术 [J]. 现代雷达, 2004, 26(3):4–7.

102. 黄美荣, 杨瑞娟. 基于IP 的可靠多播在雷达组网中的应用 [J]. 舰船电子对抗, 2003, 26 (5):21–25.

103. 吴巨红, 陈曾平. 无线局域网技术及其在远程目标数据传输中的应用 [J]. 计算机工程, 2003, 29(2):127–128.

104. 单奇. 应用于雷达组网的数据接入单元设计 [J]. 电子工程师, 2004, 30(6):6–9.

105. 张宏荣, 夏传浩, 刘进平. 情报雷达组网中工作站间的实时数据通讯 [J]. 现代电子, 1995, 4:35–44.

106. 黄银园. 大型多基地雷达分布式控制系统设计 [J]. 现代雷达, 2005, 27(5):49–52.

107. 徐军, 李强. 分布式雷达组网模型研究 [J]. 现代雷达, 2003, 25(12):5–7.

108. 何保民.　雷达网中采用的关键技术研究　[J].　海军航空工程学院学报, 　2005, 　20 (4):432–434.

109. 黄美荣, 杨瑞娟, 肖玉芬. 雷达组网中的IP实时可靠多播 [J]. 现代雷达, 2006, 28(3):4–7.

110. 马祖长, 孙怡宁.　大规模无线传感器网络的路由协议研究　[J].　计算机工程与应用, 2004.11:165–167.

111. 谢希仁. 计算机网络(第四版) [M]. 电子工业出版社, 2003.6.

112. 张永顺, 童宁宁, 赵国庆. 雷达电子战原理 [M]. 国防工业出版社, 2006.

113. 张明友. 数字阵列雷达和软件化雷达 [M]. 电子工业出版社, 2008.2.
114. 王小谟, 张光义. 雷达与探测——信息化战争的火眼金睛 [M]. 国防工业出版社, 2008.
115. P. van Genderen and W.J.H. Meijer. Non Coherent Integration in a Medium PRF Radar [C]. IEEE International radar conference, 1995: 91–93.
116. Y. Teng, H.D. Griffiths, C.J. Baker. Netted radar sensitivity and ambiguity [J]. IET Radar Sonar Navig, 1,(6), 2007:479–486.